Divergence and Convergence of Automobile Fuel Economy Regulations

Masahiko Iguchi

Divergence and Convergence of Automobile Fuel Economy Regulations

A Comparative Analysis of EU, Japan and the US

Springer

Masahiko Iguchi
Department of International Relations
Faculty of Foreign Studies
Kyoto Sangyo University
Kyoto
Japan

ISBN 978-3-319-17499-0 ISBN 978-3-319-17500-3 (eBook)
DOI 10.1007/978-3-319-17500-3

Library of Congress Control Number: 2015935208

Springer Cham Heidelberg New York Dordrecht London

Printed on acid-free paper

Springer International Publishing AG Switzerland is part of Springer Science+Business Media (www.springer.com)

For my parents, who have been giving me enduring supports

Preface

This book was initially born out of the questions I formulated during my time as a Ph.D. student at Tokyo Institute of Technology throughout 2007–2012. As many scholars started to engage in the role of business actors in global environmental governance, discussions associating with the 'private environmental governance' became popular research foci. These discussions look at business efforts to 'green' themselves, and many address 'beyond the state' activities of business involvement in international environmental politics, i.e. how business communities establish codes of conduct or certification schemes to promote environmentally friendly products. Then I began to wonder why business sometimes chose to support stringent environmental regulation at home country and attempt to internationalize such regulatory standards, over international business networks to coordinate their efforts to green themselves? What is such business, and would it also happen in climate change issue? Which other non-state actors are crucial to promote such business supports for stringent environmental regulation at home country? Who gets, what and how, when such business supports are happening?

Year 2007 marked remarkable changes in fuel economy regulations for passenger cars in Europe, Japan and the US: while Europe and Japan progressively started to lead the regulations, the US for the first time in 20 years raised the corporate average fuel economy standards under the Energy Independence and Security Act. At the same time, the 'Bali roadmap' was adopted under the international climate negotiation taken place at the United Nations Framework Convention on Climate Change (UNFCCC) to set processes to finalizing a binding agreement at the 15th Conference of the Parties (COP 15) in 2009. Although the COP15 resulted in a major disappointment, fuel economy regulations in Europe, Japan and the US continued to evolve and the regulatory standards for 2020–2025 started to converge throughout 2009–2012.

This book reveals the mechanism of the regulatory convergence of car fuel economy regulations between Europe, Japan and the US by drawing upon constructivist theory of International Relations and literatures that focus on business competition and environmental regulations. It offers new understandings on the topic of 'cars and carbon' by:

- dealing with the emerging phenomenon of convergence of car fuel economy regulations,
- addressing the role of the business actor in pushing towards solution of climate change issue,
- proposing the new model of 'Agency with and beyond the states',
- providing rich case studies from Europe, Japan and the US.

To this end, this book is structured as follows.

Chapter 1, 'Introduction' first highlights the automobile industry and global climate change in order to provide an issue background.

Then, Chap. 2, 'Business Actors and Global Environmental Governance' shows where the proposed monograph stands in the discipline of political science, and demonstrates how this book would advance the study of business actors in global environmental governance.

Chapter 3, 'Construction of European Fuel Economy Regulations for Passenger Cars' looks at how Europe's climate policies for car CO_2 emissions have been constructed. It asks, how and why has the European Union (EU) introduced these standards, even before Japan and the United States? What factors influenced these regulations, and which actors were instrumental in the decision-making process?

Chapter 4, 'Construction of Japanese Fuel Economy Regulations for Passenger Cars' argues that Japan adopted its stringent fuel economy regulation primarily because of industry competitiveness, which are motivated by stringent environmental regulations in export markets and encouraged by its tradition of 'co-regulation' and 'corporatism' to enhance the regulations. An earlier version of this chapter appeared in Masahiko Iguchi and Karl Hillman (2012) 'The Development of Fuel Economy Regulations for Passenger Cars in Japan'. In Nilsson et al. (eds.), *Paving the Road to Sustainable Transport: Governance and innovation in low-carbon vehicles*. London: Routledge, pp. 57–69.

Chapter 5, 'Construction of the US Fuel Economy Regulations for Passenger Cars' addresses why, despite the US being the world's first country to introduce fuel economy regulations, has US fuel economy regulation been stagnant for more than 20 years? What political dynamics pushed the former Bush Administration, which had withdrawn from the Kyoto Protocol negotiation in 2001, to improve the fuel economy regulation standard? Why are the recent US fuel economy regulations now converging with the Japanese and European standards?

Chapter 6, 'Comparative Assessment' compares and contrasts fuel economy regulations that draw implications to the target for 2015 and beyond, by comparing and contrasting fuel economy regulations among three case studies.

Lastly, Chap. 7 'Conclusion' provides broader implications to theories, explores applicability of 'agency with and beyond the state' model to other sectors, and to climate governance as a whole, by answering research questions.

This book is intended for political scientists, especially scholars of global environmental politics. Experts on climate change politics may be the most interested readers, but as the study features new aspects of private environmental governance, I hope that it would serve as a source for the wider environmental study community. Also, I hope scholars of political economy who are interested in the role of the car industry, and researchers engaged in comparative research will find use for this book. The book can also serve for practitioners, including policy makers, informed activists, advisors, business community and related professionals.

Above all, I would be extremely honoured if any student, who accidentally picked this book in the university library, would be motivated to study global environmental governance after reading this book.

Kyoto
April 2015

Masahiko Iguchi

Acknowledgments

I would like to acknowledge that this book was only possible with the support of others. Although I cannot list all their name here, I would like to take the opportunity here to acknowledge some of those who have provided such support.

First of all, I would like to express the deepest gratitude to Prof. Norichika Kanie for his guidance, support, and continuous encouragement from the beginning of this research project at Tokyo Institute of Technology. He has not only taught me valuable insights into international environmental politics, but also provided me with many research opportunities and always tried to offer me a better research environment, for which I am deeply indebted.

I am also very grateful to Prof. Frank Biermann. He offered me to host a precious research opportunity at Institute for Environmental Studies (IVM), Vrije Universiteit Amsterdam from February to March 2010, gave me elaborate guidance for my dissertation, and inspired me through his grand perspectives on Earth System Governance.

I was lucky enough to work as a research fellow at the Faculty of Law, Keio University from October 2010 to March 2012, where I gained considerable advantages to improve my insights into political science. I would like to thank Prof. Toshiro Tanaka and Professor Yoshiaki Kobayashi for giving me a chance to work at Keio University. Special and big thanks go to my ex-colleagues, Taisuke Fujita and Shin Osawa for their friendships.

Knowing the state of UN negotiations was also inspiring experience that advanced the study of business in environmental politics. I would like to thank Mr. Mitsutoshi Hayakawa, an executive director of Citizens' Alliance for Saving the Atmosphere and the Earth (CASA) for enabling me to participate in several UNFCCC COP meetings with a wonderful team.

I have also had the privilege to discuss my ideas or co-author several publications with the following scholars: Masachika Suzuki, Harro van Asselt, Karl Hillman and Yuri Okubo.

As mentioned, an earlier version of Chap. 4 of this book is based on the previously published material from Routledge. I thank Routledge for permission to use the material.

Special and big thanks go to Fritz Schmuhl, who has been more than helpful to materialize this book from the beginning.

Last but not least, extra special credit should be given to Azusa Omura, who has been supporting this research project from the beginning. She has not only been generous about cooperating, but also making every second of my life beautiful.

This work is dedicated to my parents, who have been giving me enduring support. Researching and writing this book has been long and challenging, but it was a wonderful journey indeed.

Contents

Acronyms

AAM	American Automobile Manufacturers Association
ACEA	European Automobile Manufacturers' Association
APP	Asia-Pacific Partnership on Clean Development and Climate
BRICS	Brazil, Russia, India, China and South Africa
CAFE	Corporate Average Fuel Economy
CAN	Climate Action Network
CCA	US Clean Air Act
CFC	Chlorofluorocarbon
CLEPA	European Parts Manufactures' Association
CME	Coordinated Market Economy
CO_2	Carbon Dioxide
COP	Conference of Parties
DG	The Directorate-General, European Commission
EPA	US Environmental Protection Agency
EU	European Union
FSC	Forest Stewardship Council
FTA	Free Trade Agreement
g/km	Grams of CO_2 per vehicle kilometre
GCC	Global Climate Coalition
GHGs	Greenhouse Gasses
GM	General Motors
HEVs	Hybrid Electric Vehicles
ICE	Internal Combustion Engine
IR	International Relations
JAMA	Japan Automobile Manufacturers Association
KAMA	Korean Automobile Manufactures' Association
km/L	Kilometre per Litre
LEV	California's low-emission vehicle regulation
LME	Liberal Market Economy
METI	Japanese Ministry of Economy, Trade and Industry
MLIT	Japanese Ministry of Land, Infrastructure, Transport and Tourism

MOE	Japanese Ministry of the Environment
mpg	Miles per gallon
MSC	Marine Stewardship Council
MVEG	Motor Vehicle Emission Group, European Community
NGOs	Non-Governmental Organizations
NSMD	Non-State Market Driven Governance
ODS	Ozone-Depleting Substances
PPPs	Public-Private Partnerships
R&D	Research and Development
SUV	Sports Utility Vehicles
TPP	Trans-Pacific Partnership
UAW	Union of Automobile Workers, the US
UNFCCC	United Nations Framework Convention on Climate Change
US	The United States
VDA	German Car Manufactures' Association
VOC	Varieties of Capitalism
ZEVs	California's Zero-Emission Vehicles Regulation

Chapter 1
Introduction

Abstract This introductory chapter highlights automobile industry and global climate change in order to provide an issue background. It justifies why three case studies from the EU, Japan and the US are selected. Second, it describes regulatory convergence of fuel economy regulation in greater details, and the research questions and hypothesis it seek to answer in the light of the convergence of fuel economy regulations are presented. It poses following two research questions to be addressed in this book: (1) How automobile industry can transform from the position of 'dragger' to 'pusher' towards solution of climate change issue and what are driving forces behind of such transformation? (2) Despite the fact that fuel economy regulations have been developed differently in Europe, Japan and the US, why are fuel economy standards for 2020–2025 in these countries converging? What are the political dynamics behind this trend? Finally, it describes the contributions of the book.

Keywords Regulatory convergence · Automobile industry and climate change · Japanese car fuel economy regulations · European car fuel economy regulations · US car fuel economy regulations

1.1 Transformation of Business Towards Sustainability

There is no doubt that the issue of the climate change[1] has been the most contested environmental issue internationally over the last decades. The reduction of carbon dioxide (CO_2) gained many attentions from the government, civil society and businesses. The main reason is that climate change issue is not only an environmental issue, but also reduction of CO_2 emissions deeply affects countries' energy policies as well as the structure of economic activities.

---

[1]The issue of climate change refers to "a change of climate which is attributed directly or indirectly to human activity that alters the composition of global atmosphere and which is in addition to natural climate variability observed over comparable time periods" (Article 1, UNFCCC 1992).

M. Iguchi, *Divergence and Convergence of Automobile Fuel Economy Regulations*,
DOI 10.1007/978-3-319-17500-3_1

In particular, the role of business actors in international environmental politics has been gaining scholarly attentions in the discipline of International Relations (IR) in recent years (Rowlands 2001; Jacobs 1991; Hurrell and Kingsbury 1992; Clapp 1998, 2003; Usui 2002; Desombre 2005; Levy and Newell 2005; Börzel and Risse 2005; Pattberg 2005, 2007; Chan and Pattberg 2008; Pattberg and Stripple 2008; Cashore 2002; Bernstein and Cashore 2007; Cashore et al. 2007; Falkner 2001, 2003, 2005, 2008; Bäckstrand 2008). Business actors attracts such scholarly attentions because business interests are directly affected by environmental regulations, and their activities contribute the global environment both *negatively* and *positively* (Rowlands 2001). That is to say, while they affect the global environment negatively as their economic activities inevitably pollute the environment and spending huge resources to block any improvement in environmental regulations through lobbying activities (Newell and Paterson 1998; Levy and Newell 2002, 2005; Paterson 2007); at the same time, they are likely to support stringent international environmental regulation "where it provides them with a competitive advantage, by reducing the transaction costs of competitor firms that operate in countries with lower environmental standards" (Falkner 2008, p. 33). In other words, they promote the adoption of higher environmental standards in home country in order to create a global or regional level playing field (Vogel 2003).

These double characteristics of business actor made IR researchers interested in business and environmental politics haunted by a certain question: how can business actors transform from the position of 'dragger' to that of 'pusher' towards the solution of environmental degradations, and what are the driving forces behind such transformation?

Automobile industry in particular displays an interesting case for our journey to explore business involvements in international climate politics due to following reasons. Firstly, the road transport sector is responsible for a large proportion of CO_2 emissions, accounts for about 17 % of total global CO_2 emissions (IEA 2009a). Furthermore, the emissions are likely to rise with growing automobile production. Secondly, they create state's economic development and employment; thus, it is commonly argued that enhancing the international competitiveness of the automobile industry is essential to the prosperity of a state (Paterson 2007). Finally, enhancements of competitiveness of automobile industry require stringent fuel economy regulations to foster sustainable technologies that could potentially contribute to the solution to environmental degradations. The question is, then, *how automobile industry transforms from the position of 'dragger' to 'pusher' towards solution of climate change issue and what are driving forces behind of such transformation?*

To begin with, 'three stage model' of business interactions with international environmental politics proposed by Usui (2002) gives us a useful guidance to device an answer to the question: In the first stage, business actors lead blocking coalitions against the emergence of legally binding agreements (Newell and Paterson 1998; Levy and Newell 2002, 2005; Paterson 2007); Secondly, they form enforcement-oriented self-regulatory initiatives in various forms, such as through 'self-regulation'[2] (Webb 2002) or 'Public-Private Partnerships'[3] (Börzel and Risse 2005; Bäckstrand 2008; Pattberg 2005, 2007; Chan and Pattberg 2008; Pattberg and Stripple 2008; Pattberg et al. 2012); And third, they move on to performance-oriented 'beyond-compliance initiatives', such as 'Non-State Market Driven Governance'[4] (Cashore 2002; Bernstein and Cashore 2007; Cashore et al. 2007) or support for stringent international environmental regulations (Desombre 2005; Falkner 2001, 2003, 2005, 2008). Beyond-compliance initiatives here refer to any business endeavours to create private form of governance or to support higher environmental regulations, that is beyond the level of existing regulations. The mechanism of 'beyond-compliance' intiatives is what make the study of business involvement in environmental politics fascinating. This book asks the mechanism of the third model, namely, the key factors that gravitate the business position towards beyond-compliance initiatives.

In recent years, we are witnessing converging trends of stringent fuel economy regulations between major automobile manufacturing regions of Europe, Japan and the US (more descriptions in next section). Is this meaning 'beyond-compliance' initiatives are happening in automobile industry in each country and region, and hence support higher environmental standards in order to gain competitive advantage against competitors? Which societal actors are crucial to promote such interests, and why? Who governs (Dahl 2005), who gets, what when and how (Lasswell 1990) in the process of regulatory convergence?

In order to answer these questions, this book proposes new analytical model, 'Agency with and beyond the State'—the model that explains how the industry that operates globally but very much engaged with their national governments, is operating a role as 'Agency' in international environmental politics—by drawing upon constructivist theory in the discipline of IR. The constructivist theory emphasises that the interests of states are not given, but are socially constructed (Onuf 1989; Kratochwill 1989; Wendt 1987, 1992, 1999; Kazenstein 1996; Finnemore 1996; Adler 1997; Ruggie 1998; Hall 1999). These produced interests are also reproduced and consolidated into norms and institutions and thus become structures. In turn, the structure is established as 'reality', and further defines the interests of states, and therefore the behaviour of states. Hence, the theory enables

---

[2]Industry self-regulation occurs when corporations design and enforce the rules themselves (Haufler 2001). These rules are generally adopted voluntarily. Voluntary regulation may be defined as action that is "not forced by law not persuaded by financial incentives" (Jacobs 1991, p. 134).

[3]In Public-Private Partnerships, business participation for the delivery of public works and services is facilitated in order to enhance the implementation of public infrastructure and services (IISD 2011). As applied to international environmental politics, it is often formed for the purpose of enhancing implementation of governance by complementing state-based international regimes (Bäckstrand 2008; Börzel and Risse 2005; Pattberg et al. 2012).

[4]Non-State Market Driven Governance is a market-driven governance that is "designed to embed social and environmental norms in the global marketplace that derive authority directly from interested audiences, including those they seek to regulate, not from sovereign states" (Bernstein and Cashore 2007, p. 347).

us to explain following two dimensions: variations of state interests by focusing on the relationship between domestic social and legal norms, interests, and actions of actors; and how the structure re-construct or re-defined interests of actors. As applied to observe the regulatory convergence, the model helps us to reveal the patterns of state-automobile industry relationships that develop differently according to the context of each state and the region, as well as to explain why the international competition on fuel economy regulations were emerged and how it re-constructed fuel economy regulations in other countries that directed towards the regulatory convergence.

To this end, this book is structured as follows: Chap. 2, 'Business Actors and Global Environmental Governance: Agency with and beyond the State', is a literature review and describes the analytical framework of the book. It shows where the book stands in the discipline of the political science, and to demonstrate how the book would advance the study of international environmental politics. Then, it provides the insights of the Agency with and beyond state model in greater details as well as how the research will be operationalized, in order to set up the analysis to be conducted in Chaps. 3–5, the three empirical analysis chapters. Chapter 6, 'Comparative Assessment' compares and contrasts the analysis of the case studies in Europe, Japan and the US. What is found is that the regulatory convergence is born out from regulatory competition among the major automobile manufacturing nations with the rationale to enhance its competitiveness of the auto industry. Finally, Chap. 7, 'Conclusion' summarises my research and describes the implications for theories of IR and climate policy, and proposes future tasks.

Having said that, this introductory chapter first highlights automobile industry and global climate change in order to provide an issue background. It justifies why three case studies from the EU, Japan and the US are selected. Second, it describes regulatory convergence of fuel economy regulation in greater details, and the research questions and hypothesis it seek to answer in the light of the convergence of fuel economy regulations are presented. Finally, it describes the contribution of the book.

1.2 Automobile Industry and Global Climate Change

As mentioned, the road transport sector is responsible for a large proportion of CO_2 emissions. In 2009 the road transport sector accounted for about 17 % of total global CO_2 emissions (4,876.6 million tons of CO_2).[5] The US ranked first for its share in road transport CO_2 emissions (1,402.8 million tonnes, equal to about 29 % of the global share), followed by the EU (855.6 million tonnes, equal to about

[5]Other sector accounts for (all in million tonnes of CO_2): 11,827.1 from electricity and heat production; 1,464.1 from other energy industry own use; 5,870.9 from manufacturing industries and construction; and 3,293.4 from other sectors including residential sector.

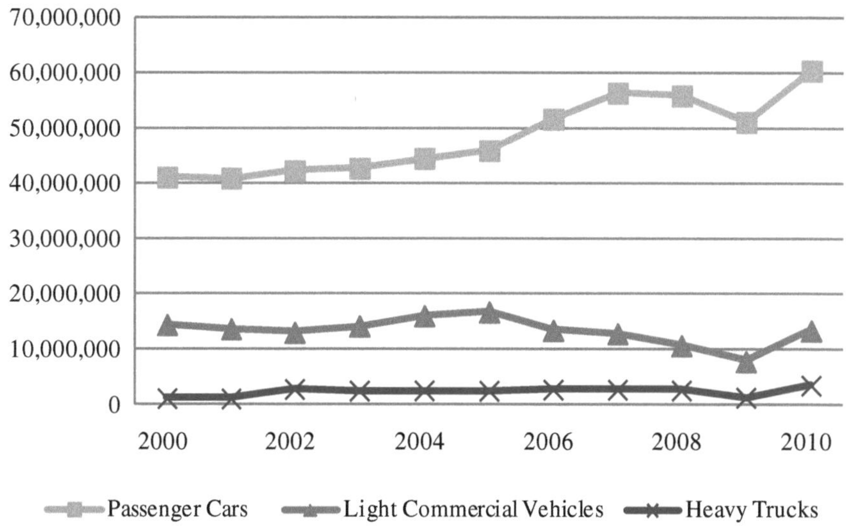

Fig. 1.1 Total numbers of vehicles produced, 2000–2010. *Source* created by the author based on OICA (2001, 2002, 2003, 2004, 2005, 2006, 2007b, 2008b, 2009b, 2010b, 2011)

18 %), China (366.5 million tonnes equals to about 7.5 %), Japan (198.2 million tonnes, equal to about 4 %), Russia (136.6 million tonnes) and India (134.1 million tonnes, equals to about 2.8 %) (IEA 2009b). Furthermore, the emissions from the road transport sector are likely to rise with growing automobile production. Figure 1.1 shows the increase of vehicles produced since 2000–2010.

Secondly, the global economic significance of the automobile industry gives it the potential political power to influence towards the low-carbon society. Global automobile production is dominated by four main regions of the world—East Asia (China and Japan), America and Europe, as shown in Table 1.1. The structural landscape of global automobile manufacturing has changed dramatically since 2009, with China becoming the biggest automotive producer in the world, producing almost as twice as much as Japan, which dropped its position as the second biggest automotive producer in the world. Still, if we focus on global automobile production in terms of individual automotive manufacturers, Japanese manufacturers (Toyota, Honda, Nissan and Suzuki), European manufacturers (Volkswagen, PSA, Fiat and Renault), and US manufacturers (General Motors, Ford and Chrysler) dominate the global automobile production market, as shown in Table 1.2. This means that, even though China has become the biggest producer in the world, automobile manufacturers from Europe, Japan and the US dominate most of the domestic production in China—for example, Chinese passenger car market is dominated by General Motors (17.8 %), Volkswagen (14.6 %), Hyundai-Kia (8 %), Nissan (7.5 %), Toyota (6.7 %), Honda (5.7 %), Ford (4.4 %), Chery (4.1 %), Geely (2.2 %) and others (29 %) (Business Insider 2010). In terms of automobile sales, although Chinese market is emerging as one of the largest automotive markets in

Table 1.1 Global automobile production share by Country, 2007–2010

Rank	Year			
	2007	2008	2009	2010
	Country (share)	Country (share)	Country (share)	Country (share)
1	Japan (15 %)	Japan (16 %)	China (22 %)	China (24 %)
2	USA (15 %)	China (13 %)	Japan (12 %)	Japan (12 %)
3	China (12 %)	USA (12 %)	USA (9 %)	USA (10 %)
4	Germany (8 %)	Germany (9 %)	Germany (8 %)	Germany (6 %)
5	S.Korea (6 %)	S.Korea (5 %)	S.Korea (6 %)	S.Korea (5 %)
6	France (4 %)	Brazil (5 %)	Brazil (5 %)	India (5 %)
7	Brazil (4 %)	France (4 %)	India (4 %)	Brazil (4 %)
8	Spain (4 %)	Spain (4 %)	Spain (4 %)	Spain (3 %)
9	Canada (4 %)	India (3 %)	France (3 %)	France (3 %)
10	India (3 %)	Canada (3 %)	Mexico (3 %)	Canada (3 %)

Source created by the author based on OICA (2007a, 2008a, 2009a, 2010a)

Table 1.2 Global automobile production share by manufactures, 2007–2010

Rank	Year			
	2007	2008	2009	2010
	Group (share)	Group (share)	Group (share)	Group (share)
1	GM (13 %)	TOYOTA (13 %)	TOYOTA (12 %)	TOYOTA (11 %)
2	TOYOTA (12 %)	GM (12 %)	GM (11 %)	GM (11 %)
3	VOLKSWAGEN (9 %)	VOLKSWAGEN (9 %)	VOLKSWAGEN (10 %)	VOLKSWAGEN (9 %)
4	FORD (9 %)	FORD (8 %)	FORD (8 %)	HYUNDAI (7 %)
5	HONDA (5 %)	HONDA (6 %)	HYUNDAI (8 %)	FORD (6 %)
6	PSA (5 %)	NISSAN (5 %)	PSA (5 %)	NISSAN (5 %)
7	NISSAN (5 %)	PSA (5 %)	HONDA (5 %)	HONDA (5 %)
8	FIAT (4 %)	HYUNDAI (4 %)	NISSAN (5 %)	PSA (5 %)
9	RENAULT (4 %)	SUZUKI (4 %)	FIAT (4 %)	SUZUKI (4 %)
10	HYUNDAI (4 %)	FIAT (4 %)	SUZUKI (4 %)	RENAULT (3 %)

Source created by the author based on OICA (2008b, 2009b, 2010b, 2011)

the world,[6] however, the market itself is smaller than the total number of passenger car sales in Europe, Japan and the US which altogether account for more than 20 million per year (ICCT 2011; RITA n.d.; JADA n.d.). Furthermore, the automobile manufacturers from Europe, Japan and the US dominate most of the sales

[6]The total numbers of automotive sales were 5.76 million in 2005; 7.22 million in 2006; 8.79 million in 2007; 9.38 million in 2008; and 13.5 million in 2009 (China Daily 2010).

in the Chinese automobile market, altogether accounting for about half of the total market sales (Business insider 2010).[7]

Based on these backgrounds, if we are to observe changes towards more sustainable road transportation sector practices at the global level, the primary focus should be placed on Europe, Japan and the US and how they could be the potential drivers to bring about such changes.

1.3 Research Puzzle and Hypothesis: Regulatory Convergence of Fuel Economy Regulation

To repeat, the main challenge of this book is to investigate how automobile industry can transform from the position of 'dragger' to 'pusher' towards solution of climate change issue and to identify driving forces behind of such transformation. In other words, what could foster endeavours of automobile industry to support higher environmental regulations, that is beyond the level of existing regulations? What are the key factors that gravitate the business position towards beyond-compliance initiatives?

Recently, we are witnessing converging trends of fuel economy regulations[8] among Europe, Japan and the US. Traditionally, Japan and Europe have had the most stringent fuel economy regulations in the world: Japan introduced 125 grams of CO_2 per vehicle kilometre (g/km) by 2015 in 2007, and 105 g/km by 2020 in 2011; Europe set a 120 g/km target in 2009 and is currently moving to set a mandatory 95 g/km by 2020. The changes occurred when President Obama announced that he would improve US fuel economy regulations to 103 g/km by 2025 (ICCT 2014). Figure 1.2 shows the recent regulatory convergence of fuel economy regulation between Europe, Japan and the US.

Is the regulatory convergence meaning automobile industry have already transformed itself from the position from the dragger to the pusher? If so, why? Who and what cultivated such transformation and how? Therefore, the second question set in this book is, *despite the fact that fuel economy regulations have been developed differently in Europe, Japan and the US, why are fuel economy standards for 2020–2025 in these countries converging? What are the political dynamics behind this trend?* Revealing these factors would then contribute to the study of the role of business actors in international environmental politics in the discipline of IR.

---

[7]Passenger car market share in Chinese market in 2009 are as follows: GM (17.8 %), Volkswagen (14.6 %), Hyundai-Kia (8 %), Nissan (7.5 %), Toyota (6.7 %), Honda (5.7 %), Ford (4.4 %), Chery (4.1 %), Geely (2.2 %) and others (29 %).

[8]According to Kerr (1983, p.3), convergence refers to "the tendency of societies to grow more alike, to develop similarities in structures process and performances". The term regulatory convergence is then about "growing similarity of institutional frameworks, policy approaches and outcomes in the field of regulatory politics" (Falkner and Gupta 2009, p. 115).

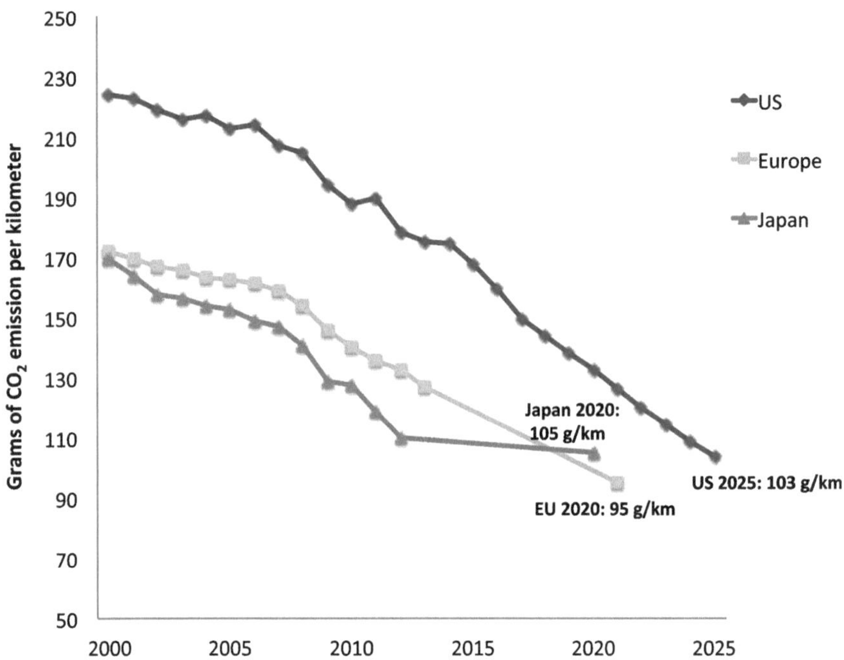

Fig. 1.2 Converged trend of fuel economy between Europe, Japan and the US. *Source* created by author based on ICCT (2014)

To answer to this research question, this book hypothesize the mechanism of regulatory convergence is the result of following three possible drivers. First hypothesis relates to an assumption that governments harmonize national policies through negotiations resulted in the regulatory convergence (Simmons 2001; Singer 2004). This hypothesis is based on an assumption that these countries and the region coordinate their policies to regulate CO_2 emissions from road transport sector, either through the United Nations Framework Convention on Climate Change (UNFCCC) or other negotiation fora that exist outside of the United Nations (UN) system. If this is happening through the UNFCCC negotiation, then it suggests that the UN system actually impacted its Parties' national climate policy. On the other hand, if Europe, Japan and the US coordinate their policy measures through non-UN negotiation fora, then it lead us to question why such fora are gaining importance.

Second hypothesis draws on what Vogel argued 'California effect', namely global competitive pressures that drive regulatory competition to adjust their national policies (Vogel 1997). According to Vogel (1997, pp. 561–562), California effect implies "[p]olitical jurisdiction which have developed stricter product standards often force foreign procedures in nations with weaker domestic standards either to design products that meet those standards, since otherwise they will be denied access to markets. This, in turn, encouraged those producers to make the investments required to produce these new products as efficiently as possible. Moreover, having made these initial investments, they now have a stake in encouraging their home markets to strengthen their standards as well, in part because their exports already meeting those standards". Hence, regulatory convergence may emerge out of exertions of automobile industry to stay competitive in foreign market where environmental standards are more stringent than home country. The automotive industry, in turn, may either actively support to raise the regulatory standard of home country, or accept stringent government regulation of home country in order to gain a competitive advantage against competitors.

Third hypothesis sketches voluntary adjustments by political actors through policy diffusion and learning (Busch and Jorgens 2005; Levi-Faur 2005). Political actors may include automobile industry, environmental Non-Governmental Organizations (NGOs) and other actors that are related the issue of fuel economy regulations. Policy diffusion and learning may be brought by concerned transnational coalitions and networks (Keck and Sikkink 1998). The argument here is that these networks influence behavior of states by excreting pressures to them: NGOs in one country appeal NGOs in other country through its network; these NGOs in turn pressure their own government to pressure other country to influence policy (so-called 'boomerang effect').[9] As applied, is the regulatory convergence resulted by boomerang effect by environmental NGOs? Or, is it business network that harmonize fuel economy of cars and support harmonious regulatory standards in each government in order to reduce trade barriers?

1.4 Research Objectives: Revealing the Dynamics of Cars and CO_2

The objective of this book is to twofold. The first research objective is to explain why fuel economy standards for 2020–2025 in Europe, Japan, and the US are converging. Specifically, it aims to answer how and why has the EU introduced the highest fuel economy regulations? (Chap. 3). How and why Japan adopted one of the highest fuel economy standards in the world, along with the EU? (Chap. 4). Finally, why, despite the US being the world's first country to introduce fuel economy regulations, has been stagnant for more than 20 years? Why are recent US fuel economy regulations now converging with the Japanese and European standards? (Chap. 5).

[9]Keck and Sikkink (1998) displayed how networked relationship of civil society transforms and challenges conceptions of national sovereignty. Drawing upon three case studies (human rights advocacy networks in Latin America, environmental advocacy networks, and transnational networks on violence against woman), they argued that these networks influence behavior of states and international organization by excreting pressures. For instance, in the case of conservation of Brazilian Amazon, the environmental advocacy network pressured the World Bank to transform its commitment on sustainable develop into action.

The second research objective relates to the core questions of business actor approach in the disciple of international environmental politics, namely, to reveal how automobile industry can transform from the position of 'dragger' to 'pusher' towards solution of climate change issue and to address driving forces behind of such transformation.

In order to explain these two research objectives, this book focuses on following variables that *cause* regulatory convergence of fuel economy standards and factors that transforms business position to the pusher: the 'Motive' of the enhancement of fuel economy regulations; the 'Competitiveness issues' that how enhancing industry competitiveness affect the regulations; the 'Decision-making process' of each regulation that backed up to materialize stringent targets; the 'Business positions' towards the stringent standards; 'NGOs role' that change behaviours of government and business to support for the stringent targets; and 'Critical juncture' that directed towards stringent fuel economy regulations.

By addressing these two aims, this book advances the study of fuel economy regulation for passenger automobiles by following three fronts. First, by revealing the logic of regulatory convergence on car fuel economy among developed countries, this book adds new insights to the discussion of 'race to the bottom or to the top' over regulatory competition and convergence (Drezner 2001; Holzinger and Knill 2004; Janicke and Jacob 2004; Prakash and Potoski 2006; Saikawa 2013; Scharpf 1997). Conventionally, the 'race to the bottom' of the regulatory convergence suggests that such convergence is based on the lowest common denominator because states are more likely to gravitate toward policies of the most laissez-fair country (Drezner 2001, p. 59). On contrary, the 'race to the top' of the regulatory convergence suggests countries compete over the stringent environmental standards for the sake of enhancing industry competitiveness. I argue that the regulatory convergence associating with fuel economy regulation among major automobile manufacturing nations are based on the race to the top. Moreover, I emphasize that what guides such regulatory competition is based on the normative consideration to stay competitive in the international automobile market.

Second, the findings of this book would make practical contributions towards the solution of climate change issue. This book demonstrates how regulatory convergence that are born out of competition among major automobile manufacturing nations, can be the 'de facto standard' in order to reduce CO_2 from the road transport sector. I argue regulatory convergence among Europe, Japan and the US becomes the 'de facto standard' and place its influence over newly emerging countries such as China and India. Altogether, Japanese, European and the US automobile manufacturers dominate global vehicle production, with the total numbers of passenger automobile market sales in these three regions totalling more than 20 million throughout 2007–2010, out of 60 million passenger automobiles produced globally. The total number of passenger car sales in Europe, Japan and the US is summarized in Table 1.3. If China and India export their automobile products to those countries, they would have to satisfy the regulatory standards among those major automobile manufacturing countries and the region. In other words, the regularity convergence among major automobile manufacturing markets work as

Table 1.3 Number of passenger car sales in Europe, Japan, and the US, 2007–2010

Year/Region	EU27	Japan	US	Total
2007	15,596,339	1,299,168	7,562,334	24,457,841
2008	14,338,100	1,250,987	6,769,107	22,35,8194
2009	14,091,605	1,160,175	5,400,890	20,652,670
2010	13,305,479	1,419,909	5,635,433	20,360,821

Source created by author based on ICCT (2011), RITA (n.d.), JADA (n.d.)

the de facto standard that influences the behaviour of all states that produce automobiles, regardless of the divisions of North and South, developed or developing countries. These findings would have a great impact on the study of global climate governance.

Finally, the research outcomes gained from this study have potential applicability to sectors where there are no effective international institutions to mitigate CO_2 emissions. For instance, sectors such as electric generation share large amounts of CO_2 emissions, but there is no effective international institution that mitigates CO_2 emissions from this sector. Therefore, this research opens up new research opportunities that can investigate to what extent de facto standards of low-carbon technology in different sectors can work as informal but effective institutions, in the context of fragmented governance.[10]

References

Abbott K (2012) The transnational regime complex for climate change. Environ Plann C Gov Policy 30(4):571–590

Adler E (1997) Seizing the middle ground. Eur J Int Relat 3:319–364

Andonova LB (2010) Public-private partnerships for the earth : politics and patterns of hybrid. Glob Environ Polit 10(2):25–53

Bäckstrand K (2008) Accountability of networked climate governance: the rise of transnational climate partnerships. Glob Environ Polit 8(3):74–102

Bernstein S, Cashore B (2007) Can non-state global governance be legitimate? An analytical framework. Regul Gov 1(4):347–371

Biermann F et al (2009) The fragmentation of global governance architectures: a framework for analysis. Glob Environ Polit 9(4):14–40

Biermann F et al (2010) Earth system governance: a research framework. Int Environ Agreements Polit Law Econ 10(4):277–298

[10]Fragmented governance implies an increasing diversity in arrangements that aim to address (parts of) the problem involving not only nation states, but also non-state actors including networks of scientists, civil society and business (Jagers and Stripple 2003; Biermann et al. 2009; Bäckstrand 2008; Pattberg and Stripple 2008; Biermann et al. 2010; Bulkeley and Newell 2010; Hoffman 2011; Keohane and Victor 2011; Zelli 2011; Zelli and van Asselt 2010; Abbott 2012; Andonova 2010; Okereke et al. 2009).

Börzel TA, Risse T (2005) Public-private partnerships. Effective and legitimate tools of transnational governance? In: Edgar G, Louis WP (eds), Complex sovereignty: on the reconstitution of political authority in the 21st century. University of Toronto Press, Ontario, pp 195–216

Bulkeley H, Newell P (2010) Governing climate change. Routledge, London

Busch PO, Jorgens H (2005) The international sources of policy convergence: Explaining the spread of environmental policy innovations. J Eur Public Policy 12:1–25

Business Insider (2010) Beijing's new car restrictions are going to devastate the auto industry in 2011. Retrieved 07 Aug 2014, from http://articles.businessinsider.com/2010-12-28/markets/30062006_1_auto-sales-beijing-municipal-commission-auto-industry

Cashore B (2002) Legitimacy and the privatization of environmental governance: how non state market-driven (NSMD) governance systems gain rule making authority. Governance 15 (4):503–529

Cashore B et al (2007) Revising theories of non-state market driven (NSMD) governance: lessons from the finnish forest certification experience. Glob Environ Polit 7(1):1–44

Chan S, Pattberg P (2008) Private rule-making and the politics of accountability: analyzing global forest governance. Glob Environ Polit 8(3):103–121

China Daily (2010) Chinese auto market takes over US as world's largest. Retrieved 07 Aug 2014, from http://www.chinadaily.com.cn/china/2010-01/09/content_9291861.htm

Clapp J (1998) The privatization of global environmental governance: ISO 14000 and the developing world. Glob Gov 4:295–316

Clapp J (2003) Transnational corporate interests and global environmental governance: negotiating rules for agricultural biotechnology and chemicals. Environ Polit 12(4):1–23

Dahl RA (2005) Who governs?: democracy and power in an American City, 2nd edn. Yale University Press, New Heaven

Desombre E (2005) Fishing under flags of convenience: using market power to increase participation in international. Glob Environ Polit 5(4):73–94

Drezner DW (2001) Globalization and policy convergence. Int Stud Rev 3(1):58–78

Falkner R (2001) Business conflict and US policy. In: Harris PG (ed) The environment, international relations, and U.S. Foreign policy. Georgetown University Press, Washington, DC, pp 157–177

Falkner R (2003) Private environmental governance and international relations: exploring the links. Glob Environ Polit 3(2):72–87

Falkner R (2005) The business of ozone layer protection: corporate power in regime evolution. In: Levy DL, Newell PJ (eds) The business of global environmental governance. MIT Press, Cambridge, pp 105–134

Falkner R (2008) Business power and conflict in international environmental politics. Palgrave McMillan, Basingstoke

Falkner R, Gupta A (2009) The limits of regulatory convergence: globalization and GMO politics in the south. Int Environ Agreements Polit Law Econ 9(2):113–133

Finnemore M (1996) National interests in international society. Cornell University Press, New York

Hall RB (1999) National collective identity: social constructs and international systems. Columbia University Press, New York

Haufler V (2001) Private sector: industry self-regulation in a global economy. Carnegie Endowment for International Peace, Washington DC

Hoffman MJ (2011) Climate governance at the crossroads: experimenting with a global response after kyoto. Oxford University Press, Oxford

Holzinger K, Knill C (2004) Competition and cooperation in environmental policy: individual and interaction effects. J Public Policy 24:25–47

Hurrell A, Kingsbury B (1992) The international politics of the environment: actors, interests, and institution. Clarendon Press, Oxford, pp 313–353

ICCT (2011) European vehicle market statistics. Washington DC

ICCT (2014) Global passenger car fuel economy and/or greenhouse gas emissions standards—May 2014 update. Washington DC

IEA (2009a) CO_2 Emissions from fuel combustion highlights -2010 edn. Paris

IEA (2009b) Transport, energy and CO_2: moving towards sustainability. OECD Publishing, Paris

IISD (2011) Sustainable development: is there a role for public–private partnerships? Canada, Manitoba, pp 1–7

Jacobs M (1991) The green economy: environment, sustainable development and the politics of the future. Pluto, London

JADA (n.d.) The number of sales of passenger vehicles in Japan. Retrieved 04 Aug 2014, from http://www.jada.or.jp/contents/data/type/index02.php

Jagers SC, Stripple J (2003) Climate governance beyond the state. Glob Gov. A Rev Multilateralism Int Organ 9(3):385–399

Janicke M, Jacob K (2004) Lead markets for environmental innovations: a new role for the nation state. Glob Environ Polit 4(1):29–46

Katzenstein P (1996) Cultural norms and national security: police and military in postwar Japan. Cornell University Press, Ithaca

Keck M, Sikkink K (1998) Activists beyond borders: advocacy networks in international politics. Cornell University Press, New York

Keohane RO, Victor DG (2011) The regime complex for climate change. Perspect Polit 9(1):7–23

Kerr C (1983) The future of industrial societies: convergence or continuing diversity?. Harvard University Press, Cambridge

Kratochwil F (1989) Rules, norms and decisions. Cambridge University Press, Cambridge

Lasswell HD (1990) Politics who gets what, when and how. Peter Smith Publisher, Gloucester

Levy D, Newell P (2002) Business strategy and international environmental governance: toward a neo-gramscian synthesis. Glob Environ Polit 2(4):84–101

Levy D, Newell P (2005) The business of global environmental governance. MIT Press, Cambridge

Levi-Faur D (2005) The global diffusion of regulatory capitalism. Ann Am Acad Polit Soc Sci 598:12–32

Newell P, Paterson M (1998) A climate for business: global warming, the state and capital. Rev Int Polit Econ 5(4):603–679

OICA (2001) World motor vehicle production by manufacturer World ranking 2000. paris

OICA (2002) World motor vehicle production by manufacturer World ranking 2001. Paris

OICA (2003) World motor vehicle production by manufacturer World ranking 2002. Paris

OICA (2004) World motor vehicle production by manufacturer World ranking 2003. Paris

OICA (2005) World motor vehicle production by manufacturer World ranking 2004

OICA (2006) World motor vehicle production by manufacturer World ranking 2005. Paris

OICA (2007a) Production statistics. Retrieved from http://oica.net/category/production-statistics/2007-statistics/

OICA (2007b) World motor vehicle production by manufacturer World ranking 2006. Paris

OICA (2008a) Production statistics. Retrieved from http://oica.net/category/production-statistics/2008-statistics/

OICA (2008b) World motor vehicle production by manufacturer World ranking 2007. Paris

OICA (2009a) Production statistics. Retrieved from http://oica.net/category/production-statistics/2009-statistics/

OICA (2009b) World motor vehicle production by manufacturer World ranking 2008. Paris

OICA (2010a) Production statistics. Retrieved from http://oica.net/category/production-statistics/2010-statistics/

OICA (2010b) World motor vehicle production by manufacturer World ranking 2009. Paris

OICA (2011) World motor vehicle production by manufacturer World ranking 2010. Paris

Onuf N (1989) World of our making. University of South Carolina Press, Columbia

Okereke C, Bulkeley H, Scheder H (2009) Conceptualizing climate change governance beyond the international regime. Glob Environ Polit 9(1):56–76

Pattberg P (2005) What role for private rule-making in global environmental governance? Analysing the forest stewardship council (FSC). Int Environ Agreements Polit Law Econ 5 (2):175–189

Paterson M (2007) Automobile politics. Cambridge University Press, Cambridge

Pattberg P (2007) Private institutions and global governance. The new politics of environmental sustainability. Edward Elgar, Cheltenham

Pattberg P, Stripple J (2008) Beyond the public and private divide: remapping transnational climate governance in the 21st century. Int Environ Agreements Polit Law Econ 8(4):367–388

Pattberg P et al (2012) Public-private partnerships for sustainable development. Edward Elgar, Cheltenham

Prakash A, Potoski M (2006) Racing to the bottom? Trade, environmental governance, and ISO 14001. Am J Polit Sci 50(2):350–364

RITA (n.d.) U.S. sales or deliveries of new aircraft, vehicles, vessels, and other conveyances. Retrieved 07 Aug 2014, from http://www.bts.gov/publications/national_transportation_statistics/html/table_01_12.html

Rowlands I (2001) Transnational corporations and global environmental politics. In: Wallace W, Josselin D (eds) Non-state actors in world politics. Palgrave McMillan, Basingstoke, pp 133–149

Ruggie J (1998) Constructing the world polity. Routledge, London

Saikawa E (2013) Policy diffusion of emission standards—is there a race to the top? World Polit 65:1–33

Scharpf FW (1997) Introduction: the problem-solving capacity of multi- level governance. J Eur Public Policy 4(4):520–538

Simmons BA (2001) The international politics of harmonization: the case of capital market regulation. Int Org 55:589–620

Singer DA (2004) Capital rules: the domestic politics of international regulatory harmonization. Int Org 58:531–565

UNFCCC (1992) United Nations Framework Convention on Climate Change

Usui M (2002) The role of private business in international environmental governance. Working Paper IAS–CGP Project on International Environmental Governance

Vogel D (1997) Trading up and governing across: transnational governance and environmental protection. J Eur Public Policy 4(4):556–571

Vogel D (2003) The hare and the tortoise revisited: the new politics of consumer and environmental regulation in Europe. Br J Polit Sci 33(4):557–580

Webb K (2002) Voluntary codes: private governance, the public interest and innovation. Carlton University, Otawa

Wendt A (1987) The agent-structure problem in international relations theory. Int Organ 41(3)

Wendt A (1992) Anarchy is what states make of it: the social construction of power politics. Int Organ 46(2)

Wendt A (1999) Social theory of international politics. Cambridge University Press, Cambridge

Zelli F (2011) The fragmentation of the global climate governance architecture. Wiley Int Rev Clim Change 2(2):255–270

Zelli F, van Asselt H (2010) The overlap between the UN climate regime and the world trade organization: lessons for post-2012 climate governance. In: Biermann F, Frank P, Philipp Z (ed) Global climate governance beyond 2012: architecture, agency and adaptation. Cambridge University Press, Cambridge, pp. 79–96

Chapter 2
Business Actors in Global Environmental Governance

Abstract The purpose of this chapter is twofold: to show where the book stands in the discipline of the political science, and to demonstrate how it would advance the study of business actors in global environmental governance. To begin with, this chapter firstly classifies existent literatures on business actors approach in the discipline of global environmental governance into following four categories: business self-regulations, public-private partnerships, non-state market driven governance and business conflict school. Second, based on the first point, a constructivist perspective on business actors and environmental governance is introduced. Finally, given the insights of the constructivist theory, it explains how the case of automobile industry would add new insights on extant approaches.

Keywords Business actors · Global environmental governance · Agency beyond the state · Agency with and beyond the state · Constructivism

2.1 Business Actors in Global Environmental Governance: A Classification of Business Involvement

As the name 'International Relations' suggests, the principal analytical units have been nation-states. There has been a significant amount of writing within the IR discipline on the international politics of the environment in recent years. In particular, much of the works are done in line with the regime theory[1] mostly outlined

[1]Liberal institutionalist position associates with so-called 'Neo-Liberal Institutionalism' in International Relations. The core assumption of the theory is placed on the role of international institutions in international politics. By drawing upon the Prisoner's Dilemma in game theory, it argues that the key to solve the Dilemma is to convince and shift the perception of the other parties that it

M. Iguchi, *Divergence and Convergence of Automobile Fuel Economy Regulations*,
DOI 10.1007/978-3-319-17500-3_2

by Keohane (Keohane 1988; Keohane and Axelrod 1985; Keohane and Levy 1996; Keohane and Nye 2000). This is not surprising, given that certain forms of environmental degradation, such as the issues of climate change and ozone depletion, cannot be solved purely at the local level, but their solution requires global cooperation since they affect the planet as a whole (Doyle and McEachern 1998; Connelly et al. 2002). With the strong emphasis on international regimes,[2] international cooperation, and the ordering and management of the interstate system, regime theory continues to enjoy a privileged status within the study of international politics of the environment (Newell 2005).

As applied to the study of global governance, Young (1999) has suggested that 'governance without government' is possible by international regimes. Young conceived global governance as consisting primarily of interstate cooperation or interdependence, and he therefore analysed as a collective actions problems. The regimes, or sets of roles, rules and relationships that focus on specific issue areas could provide governance under the condition of anarchy. The primary components of these regimes are nation-states, and therefore its focus was placed on international regimes. It follows that non-state actors are less important in the process of governance, and they are primary understood as a factor that enhances the effectiveness of interstate patterns of governance.

In the recent study associating with global governance suggests that global governance is about "regulatory mechanisms in a sphere of activity which function effectively even though they are not endowed with formal authority" (Rosenau 1995, p. 5). This means that not only nation-states, but also non-state actors are increasingly taking important roles in global governance. The section below discusses how various non-state actors are gaining importance in global environmental governance.

In the context of environmental politics, the identification of domestic factors that shape the interest of state actors is essential for a better understanding of the topic. Rather than structural constraints, domestic political pressures have large

(Footnote 1 continued)
is possible to gain mutual benefits achieved by long term cooperation, rather than the imaginary benefits that could be gained by the act of Defection. To put simply, it is needed that Prisoner's Dilemma to be played over time. To ensure the game to be played over time is to establish international regimes. Regimes facilitate cooperation because for example, they do change patterns of transaction costs and provide information to participants, so that uncertainty is reduced. Thus regimes facilitate cooperation because they disseminate information, monitor behaviour and therefore help to prevent cheating. Moreover, regimes facilitate cooperation among states because they: facilitate issue inter-linkage; increase number of interaction; help states to find partners to cooperate because regimes allow reputation to be developed and this influence states' future potential to do so (Keohane and Axelrod 1985).

[2]Regimes refers to a set of explicit of implicit "principles, norms, and decision making procedures around which actor expectations converge in a given issue-area" (Krasner 1983).

impacts over state positions and negotiation preferences in foreign climate change policies. For instance, business actors, the scientific community and environmental NGOs, are actors that shape the interest of state actors in the context of the international politics of the environment. Thus, the examination of the extent of their influence over climate change politics is crucial.

This demand of the further investigation on the role of the non-state actor in international environmental politics led to the new strand of research area in the discipline of IR, which is labelled as 'Agency beyond the State' (Biermann et al. 2009; Dellas et al. 2011). The concept of Agency and actor differs. Actor can refer to any party that takes action, whereas Agency can be conceptualized as any actor who possesses the ability to prescribe behaviour and to obtain the consent of the governed. In this context, they may not only "lobbying and advising national governments in the creation and implementation of rules", but also they "substantially participate in/or set their own rules related to the interactions between humans and their natural environment" (Biermann et al. 2009, pp. 37–43). The key focus is placed on the need of the further investigations on whether nation-states can fulfil their core functions under the pressure of earth system transformation,[3] and to what extent non-state actors are filling new governance demands. 'Agency' may include various non-state actors, such as scientific community or what Haas called the 'epistemic community' (Haas 1992), 'global civil society' (Wapner 1995; Lipschutz 2004) and business actors.

This book sheds light on the business actor in the context of environmental politics. The interests of business actors are directly affected by environmental regulation. Most notably, their activities contribute the global environment both *positively* and *negatively* (Rowlands 2001). Needless to say, they affect the global environment negatively as their economic activities inevitably pollute the environment and thus create various environmental degradations. At the same time, increasing investments on research and development on sustainable technologies could potentially contribute to the solution to environmental degradations.

The involvement of business actors in the international politics of environment is not new. The business involvements in the United Nations environmental governance system can be traced back to 1977 when business actors launched a global applicable voluntary code of conduct for themselves as the UN Centre on Transnational Corporations (UNCTC).[4] Another example can be drawn from the

[3]Earth system transformation is marked by "persistent uncertainty regarding the causes of global environmental change, its impacts, the interlinkage of various causes and response options, and the effects of possible response options….Uncertainty hence poses particular governance challenges. It requires governance to be stable over decades and centuries to withstand sudden changes of earth system parameters (or changes in out knowledge about these parameters), but also to be flexible enough to adapt to changes within the larger stable framework. Governance must be oriented towards the long term, but must also provide solutions for the near future" (Biermann 2007, pp. 329–330).

[4]The UNCTC was established in 1972 as the focal point, for all matters related to transnational corporations and foreign direct investment within the United Nations system. In 1993, the UNCTC was transferred to the Division on Investment, Technology and Enterprise Development of the

establishment of the World Business Council for Sustainable Development (WBCSD) in 1995[5] and Business Action for Sustainable Development (BASD) in 2002.[6] Further, the UN and the business actors made a pact called 'The Global Compact' in 2000 that set ten universally accepted principles for corporations, including environmental matters such as supporting a precautionary approach, promoting greater environmental responsibility and encourage the development of environmental friendly technologies. This pact primarily aimed for the business actors to behave in socially and environmentally sound ways (Clapp 2005).

Given the growing participation of business actors in making of global environmental governance, focus on the activities of business actors is increasingly important (Rowlands 2001; Jacobs 1991; Hurrell and Kingsbury 1992; Clapp 1998, 2003; Usui 2002; Desombre 2005; Levy and Newell 2005; Börzel and Risse 2005; Pattberg 2005, 2007; Chan and Pattberg 2008; Jagers and Stripple 2003; Pattberg and Stripple 2008; Cashore 2002; Bernstein and Cashore 2007; Cashore 2002; Falkner 2001, 2003, 2005, 2008; Bäckstrand 2008). Consequently, the concept of global environmental governance is changing from intergovernmental bargaining arrangements to a more dynamic, complex form of governance where business actors are actively involved.

Any theory of International Relations that focus is ultimately placed on the role of nation-states cannot account for these dynamics, as they treat the interest of states as 'given'. Therefore, the study of business and industry interests adds an important dimension to our understanding of international environmental agreements and helps to explain why some states are more active than other states in environmental policy makings. In particular, the fundamental question that lies in the growing scholarly attentions is, how industry can transform from the position of 'dragger' to that of 'pusher' in the area of global environment rule making and what could the driving force behind of such transformation.

Various theories challenge the question of changing business behaviours towards sustainability from different angles. It is important to briefly review these theoretical developments. To begin with, scholars who are labelled as the 'Neo-Gramscian strand' argue that since corporations are central to capital accumulation in each state, they possess structural influences over the state's environmental decision making (Levy and Newell 2002, 2005; Newell and Paterson 1998). The starting point of this argument is that the role of the states is to maintain and advance the general interest

(Footnote 4 continued)
United Nations Conference on Trade and Development United Nations Conference on Trade and Development (UNCTAD) (UNCTAD) in Geneva.

[5]WBCSD was originally established in 1990 as Business Council for Sustainable Development among 48 business leaders to represent the voice of business at the 1992 Earth Summit in Rio. The WBCSD was formed in 1995 in order to "galvanize the global business community to create a sustainable future for business, society and the environment" (WBCSD n.d.).

[6]BASD is a joint initiative between WBCSD and International Chamber of Commerce in order to form a "comprehensive network of business organizations that have come together under one banner in the interests of sustainable development" (BASD n.d).

of the capital, which would in turn maintain the legitimacy of the state. Consequently, those who organize the process of capital accumulation would gain great structural power over state decision-making. With regard to the politics of climate change, Newell and Paterson (1998) argue that since technologies associating with the use of oil and coal has been central to the ninetieth and twentieth century capital accumulation, fossil fuel companies were conferred a great structural power over state decision-making. The changes to this structure would be trigged by a counter-hegemony, which is defined as "a creation of an alternative hegemony on the terrain of civil society in preparation for political change" (Pratt 2004, p. 332). The force of counter-hegemonic movement often rises from civil society. Drawing upon case studies from the NGO campaign for democratisation in Egypt's during 1970s, for example, Pratt (2004) claims that the NGO campaign represented part of a counter-hegemonic movement that gained wider public support led the Egyptian government to promulgate a more democratic NGO law. So too, in the area of environmental issues, the role of civil society is also gaining importance as a source of counter-hegemony that could gravitate corporate behaviour from the dragger to the pusher (Bendell 2000; Newell 2001a, b; Carroll 2007; Pearse 2010).

Second, scholars who emphasise the importance of business 'self-regulation' (Webb 2002) and 'private governance' (Pattberg 2007; Pattberg and Stripple 2008) argue for the new form of environmental governance, as opposed to the 'public governance' provided by states.[7] This private form of environmental governance can be summarised as the business efforts to create environmental and socially sound institutional arrangements, where the business actors voluntarily structure and direct their behaviours in an issue-specific area (Clapp 2005, p. 24; Falkner

[7]As applied in the case of climate change politics, there are several evidences that support business self-regulations. These supports are driven by the perceptions that "over the long term the world will have to deal with climate change, so their climate-friendly investments will pay off" (Bang et al. 2005, p. 292). Firstly, the Global Climate Coalition (GCC), which exercised strong anti-active lobbying efforts to block any international regulations of GHGs emissions during 1990s, was dismissed in early 2002. By 1997, with the growing scientific and public consensus regarding the high risks of climate change issue, a number of GCC supporters reconsidered the negative PR implications of their involvement in the group. Consequently, with the withdrawn by the BP, a numbers of major companies abandoned GCC such as American Electric Power, Dupont, Shell, Ford, Daimler Chrysler, Texaco and General Mortars (Source Watch n.d.). Secondly, instead of abandoning the GCC, some of these business actors such as BP and Shell, formed a pro-active environmental coalition called the 'Partnership for Climate Action', which aims to reduce their aggregate emissions by 15 percent from 1990 levels by 2010 using market-based mechanisms, such as by developing an internal carbon trading scheme (Bang et al. 2005, pp. 291–292). Thirdly, the 'Chicago Climate Exchange Chicago Climate Exchange' (CCX, for short) is established in June 2001. The CCX is a 'greenhouse gas emission registry, reduction and trading system for all six greenhouse gasses' where 'members make voluntary but legally binding commitment to reduce greenhouse gas emissions'. The member of CCX includes such as Ford, DuPont, and American Electric Power. It sets the goal for all members to reduce direct emissions of 4 % below a baseline period of 1998–2001, by the end of December 2006 (CCX n.d.). Fourthly, some oil companies such as BP and Shell have begun to invest in solar energy. In the case of Shell, it established the Shell International Renewables in 1998, and invested $500 million over five years in renewable energy (Levy 2005, p. 84).

2003, p. 72). This differs from the intergovernmental negotiation-based regimes in the sense that it is driven by the commercial gains to be made from product endorsement, reduced transaction costs, and access to markets. Consequently, actors are able to make faster-track decisions as well as having equal rights to representation and transparency of proceedings. Although business actors strongly resisted international environmental agreements when they were not in their interest, we have witnessed numbers of business endeavours to 'green' themselves in recent years. Recently, there are growing initiatives in business to voluntary reduce greenhouse gas emissions. For instance, in the steel industry, companies such as Alcoa, Nippon Steele Corporation and Norsk Hydro set their own emission reduction targets; and this is also evident in the cement and petroleum industries. This shift in business actors' behaviour emerged due to the huge societal and political pressures on their environmentally negative images. In reaction, business actors chose to voluntarily support pollution prevention in order to prevent damage to their corporate images (Porter and Brown 1996, pp. 64–65), which "provided a key route for firms to project their legitimacy as responsible environmental actors" (Levy and Newell 2002, p. 93).

Based on existing literatures that deal with business actors and international environmental politics, I classify four types of business involvement in global environmental governance: voluntary regulation, public-private partnerships (PPPs), non-state market driven (NSMD) governance, and business conflict based governance or internationalization of domestic politics. Figure 2.1 shows the classification of private environmental governance.

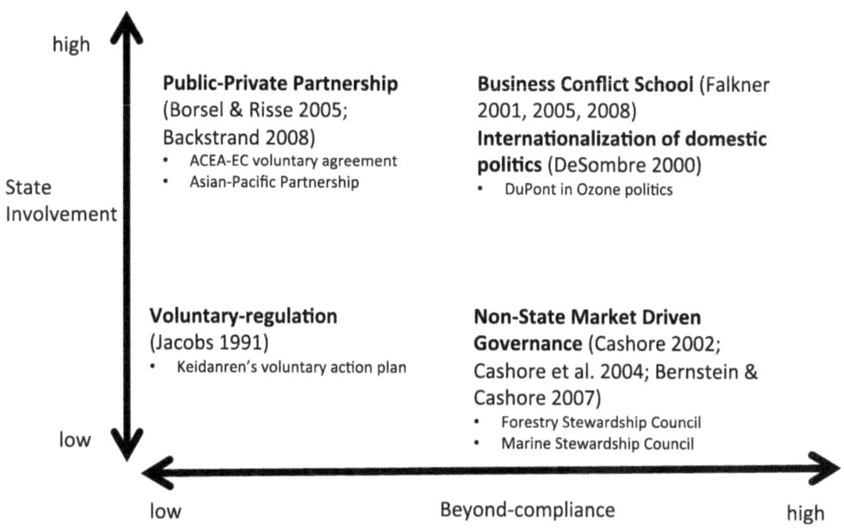

Fig. 2.1 Classification of business involvement in environmental governance. *Source* Created by author

Voluntary regulation may be defined as action that is "not forced by law not persuaded by financial incentives" (Jacobs 1991, p. 134). The Japanese Federation of Economic Organizations' (Keidanren) 'voluntary action plan', which aimed to stabilize CO_2 emissions from fuel combustion and industrial processes at 1990 level by 2010 falls into this category (Keidanren 1997). Keidanren has strongly opposed government interventions, including the use of economic instruments such as carbon taxes and emission trading. Thus, their voluntary action plan is set in order to safeguard against any governmental regulations.

The strength of PPPs in sustainable development is to enhance implementation in governance (Bäckstrand 2008; Börzel and Risse 2005). The good example of PPPs in climate change issue is Asian-Pacific Partnership on clean development and climate (APP). The APP was formed by the United States, Japan, Australia, China, India, and Korea in July 2005. In 2007, Canada joined the APP, and consequently there were seven member countries in this partnership. The aim of the APP was to pursue climate mitigation through a voluntary, non–legally binding, technology-oriented approach. Although the founder of the APP stressed that it was intended as a 'complementary' institution to the Kyoto Protocol, it has been argued that the APP was actually intended as an 'alternative' institution to the legally binding protocol (Van Asselt 2007; Christoff and Ekersley 2007; Lawrence 2007; McGee and Taplin 2006). This form of governance perhaps complements the state-based international regimes; however, given its voluntary nature, industry does not have any incentives to commit any reduction targets that are beyond the compliance.

NSMD governance is a market-driven governance that is "designed to embed social and environmental norms in the global marketplace that derive authority directly from interested audiences, including those they seek to regulate, not from sovereign states" (Bernstein and Cashore 2007, p. 347). Examples of this type of governance are the Forest Stewardship Council (FSC) and the Marine Stewardship Council (MSC). Both aim to promote responsible management of natural resources through standard setting, certification and labelling of products. For instance, the FSC had certified 116 million hectares (2.9 per cent) of forest up to 2009, by applying its principles to certify sustainable forest management (European Commission 2011), and the MSC had certified more than 100 fisheries (equal to the supply over 7 % of all the seafood we eat) around the world (WWF n.d.). In contrast to the voluntary regulation, NSMD is designed to create binding enforceable rules where the compliance mechanisms are developed over years accompanying with the market demands (Cashore 2002).

Finally, the 'business conflict school' or scholars who advocate 'internationalization of domestic politics' argues that it is the competition among corporations based on sustainable technological innovation that encourages the home state to push for more stringent international environmental regulations (Falkner 2001, 2003, 2005, 2008; Desombre 2005). In this theory, 'business competition' over the sustainable technological innovation is the fundamental driving force that changes the business behaviour. The good example would be Dupont's support for Montreal Protocol in Ozone Protection regime. As this theory has a close relevance to the argumentation of the book, detailed discussions are provided in the next section.

Having said the four typologies of business involvements in international environmental politics, this book aims to advance the study of business actor approach in this field by studying the regulatory convergence of fuel economy regulation by drawing insights from 'business conflict school' and 'internationalisation of domestic politics' approaches. It reveals that automobile industry could transform from the position of 'dragger' to 'pusher' towards solution of climate change issue through business competitions over stringent fuel economy regulations. Business competitions are motivated to create a global or regional level playing field in order to have competitive advantage over competitor firms. This, in turn, leads each government to promote higher fuel economy regulations. I argue that regulatory convergence of fuel economy regulations is born out from regulatory competition among the major automobile manufacturing nations with the rationale to enhance its competitiveness of the auto industry. Thus, this book contributes to the study of involvements of business actors in international environmental politics by revealing that business conflicts motivated by enhancing the industry competitiveness have been the central determining factors behind such transformation.

2.2 Existing Studies: Varieties of Capitalism and Environmental Policy-Making

Previous literature in this field suggests that business strategies are deeply rooted in a country's historical context as well as its institutional environment, and this is particularly the case for the automobile industry (Levy 2005; Mikler 2009). Existing research that compares fuel economy regulations between Europe (with main focus on Germany), Japan, and the US was conducted by Mikler (2006, 2007, 2008, 2009, 2010) by examining varieties of capitalism (VOC). The VOC approach assumes different institutional structures according to different countries, especially the relationships between government and industry and how they influence different patterns of institutional structure in each country (Hall and Soskice 2001).

Mikler has attempted to address the differences in state-automobile industry relationships in these nations, by classifying Japan and Europe as the Coordinated Market Economy (CME) and the US as the Liberal Market Economy (LME). Mikler argued that the CME countries are likely to have higher standards of fuel economy regulation because of the closer (cooperative) the relationship between state and industry, while, he explained, US fuel economy regulation has been stagnant because of its LME tradition.

While VOC approach has greatly advanced our understandings on the relationships between different types of Capitalist system and environmental standards, however, since its focus is placed on the *institutional arrangements* of each country, it fails to account dynamics of actor relationships that actually triggered the convergence of fuel economy regulations. Focusing solely on the domestic arrangements in each country could not adequately reveal the dynamics of automobile environmental politics.

This book, however, argues that the 'Agency with and beyond state model' can explain these trends better than the VOC approach. Drawing upon holistic constructivist approach, the model enables to broaden the scope of analysis to analyse how different interests regarding fuel economy regulations and state-automobile industry relationships have developed, how they have interacted, and how these interactions have resulted in converging fuel economy standards (more discussions in next section about holistic constructivist approach).

Furthermore, based on existing literatures that stress the close correlation between business competition and environmental policy, the book argues business competitiveness concern is the fundamental factor that have been constructing fuel economy regulations in each country. In other words, countries have constantly tried to introduce stringent fuel economy regulations in order to enhance the competitiveness of the automobile industry, to succeed in the global market, and to survive in the global market. As a result, what has emerged is a trend towards converging fuel economy standards worldwide.

2.3 A Constructivist Perspective on Business Actors and Environmental Governance

This section introduces constructivist theory of IR, explaining how it offers better explanation than other theories, and further navigates us to deeper understanding associating with regulatory convergence of fuel economy standards. There are three ontological propositions of Constructivist theory in general (Reus-Smit 2005): emphasis on the importance of normative and ideational structures; identities construct interests; and, agents and structures are mutually constituted.

On the first point, as Berger and Luckmann (1967) emphasised the importance of normative and ideational structures rather than material structures like rationalists,[8] because normative structures are thought to shape the identities and interests of actors through imagination, communication and constraint. The central claim of constructivists lies in its focus on the role of ideas as structures that constrain and shape actors' behaviour, and therefore there is an emphasis on the role of ideas and shared knowledge in the social world. The ideational structure not only has regulative effects on actors (Wendt 1987, 1992, 1999), but also has constitutive effects on actors, as structures leads actors to redefine their interests and identities in the process of interacting.

The second point, constructivists assume that identities constitute interests, in contrast to rationalists, who emphasise relative gain in world politics. Constructivists see 'identities are the basis of interests' (Wendt 1992), and focus on how

[8]'Rationalists' refers to those who adopted so-called neo-realism and/or neo-liberalism, those of which emphasize the importance of international system and treat the interest of states as given.

normative structures shape and (re)constitute the behaviours of actors. Therefore, according to constructivists, the system of state is a result of a process of internationalisation of new identities and interests.

The third point, with an influence from 'structurationism' by Giddens (1986), emphasises that ideational structures and actors co-constitute and co-determine each other's identities. Structures constitute actors in terms of their interests and identities, but structures are also produced, reproduced and altered by the discursive practices of actors. This suggests that actors can change structures through acts of social will.

Although constructivists commonly agree about these ontological assumptions in general, they emerge out of critiques of 'positivism' and neo-utilitarian IR theory, often labelled as the 'middle way' or 'new orthodoxy'[9] (Keohane 1988). Nevertheless, there are different strands of constructivism on epistemological and methodological grounds. This variation needs to be explained before we move on to introduce the explanatory variable of the theory. One way to classify such differences is between reflectivist and positivist constructivism. Reflectivists such as Adler (1997), Kratochwil (1989) and Onuf (1989) refused to adopt methods or devise frameworks for analysis, arguing that social objects are simply not describable in terms of categories of pure observation or measurement procedures. On the contrary, positivist Constructivists emphasised that systems and agents are mutually constitutive and that any divide between them is ahistorical.

To put more simply, Constructivist theory can be divided into the 'thick' and 'thin' Constructivism (Smith 1999): while the former tends to adopt normative analysis (e.g. discourse analysis) in order to observe how norms and discourses shape international society and behaviour of actors; the latter tend to focus on empirical research to explain how norms that generated through interactions among various actors shape the formations of state interests (Finnemore 1996). This book adapts to the latter approach, and focuses on the involvement of business actors in international environmental politics.

Another way of categorizing Constructivist theory is differing viewpoint on the question of 'Agent-Structure' problem, namely, whether the emphasis is placed on (1) how international structure shape and limit the behaviour of actors, or (2) how international structures are constructed as a result of interactions among actors. Reus-Smit (2005) categorizes into following three constructivists—systemic, unit-level, and holistic. Systemic constructivists, such as Wendt (1999), follow neo-realism in adopting the 'third image' perspective (Waltz 1959). The third image perspective focuses solely on interactions between unitary state actors and advocates pure systemic theorising, drawing distinctions between the domestic and international systems and ignoring the former. The limitation of this approach lies

[9]The rise of Constructivism is sometimes referred as 'the middle way' between Rationalism (a theory of International Relations that adopted positivist epistemology) and Reflectivism (other theories that reject positivism, including critical theory and postmodernism). Constructivism is also labeled as 'new orthodoxy', as it offers new ontological perspective that is different from Neo-Realism and Neo-Liberal Institutionalism.

in its narrow realm, and its inability to explain how fundamental change occurs, since it leaves out domestic factors.

Unit-level constructivism, characterized by Katzenstein (1996), focuses on the relationship between domestic social and legal norms and identities, interests, and actions of states. It draws attention to the domestic determinants of national policies, which enables it to explain variations of identity, interests and actions across states. However, on the other hand, "this form of constructivist has difficulty accounting for similarities between states, for patterns of convergence in state identity and interest" (Reus-Smit 2005, p. 200).

This book takes the third type of constructivism, often labelled as 'Holistic constructivism', characterized by Ruggie (1998) and Hall (1999). It challenges the dichotomy between international and domestic politics, and seeks to bring them together into a unified analytical perspective, focusing on a mutually constitutive relationship. Drawing upon insights of Holistic constructivism, this book sheds a light on the 'blackbox' of interest formation process of state decision-makings on car fuel economy regulations.

How domestic politics, such as institutions, the preference of societal actors, and domestic political commitments determine the expected political, economic and legal impact of international commitments have been drawing scholarly attention since 1990s (Weaver and Rockman 1993; Goldstein 1996; Gourevitch 1996; Raustiala 1997). Domestic politics matter to a great extent in international environmental politics (De Sombre 2000; Vogel 2003; Bramble and Porter 1992; Paarlberg 1996; Schreurs and Economy 1997; Sussman 2004; Underdal and Hanf 2000). International environmental agreements are commonly aims to transform domestic rules or standards, and hence influence behaviours of private actors towards sustainability. Since the fundamental objective of international environmental agreements is to regulate the activities of private actors in each country, they lobby governments accordingly. The degree of the success of international environmental agreements is largely affected by domestic contexts, because domestic institutions and political structures shape the position of each government.

What is also important is the notion of 'internationalization of domestic environmental regulations' (De Sombre 2000). It is an endeavour of states to convince other states to adopt similar regulatory standards for following two reasons. First and foremost, states acting alone cannot solve the environmental degradations such as climate change issue, and there are simply no incentives for any states to act alone. Therefore states push other states to engage into effective agreements. The second reason relates to enhancing the competitiveness of the private actors. The case from ozone depletion[10] illustrates the most prominent example. The United States was very active in pushing for Chlorofluorocarbon (CFC) reductions, both domestically and internationally. Domestically, it introduced the legislation in 1978, which banned CFCs in 'non-essential' aerosols. Internationally, it proposed

[10]Ozone depletion is an issue that the total volume of the ozone layer is destroyed by man-made ozone-depleting substances (ODS), such as halocarbon Chlorofluorocarbon (CFC).

for the global reduction of ozone-depleting substances (ODS) at the Montreal Protocol negotiation. This US leadership led the revision process of the Montreal Protocol faster and stronger (Rowlands 1995, pp. 102–122). Behind of this US leadership, there were corporate and government interests to take the 'first mover's advantage'.[11] The corporate interests appeared when the Dupont and the Alliance announced their support for international controls on CFCs in August 1986. In doing so, DuPont had decided to spend great deal of money on finding CFC alternatives in 1986 onwards. By establishing two industry programs to assess the environmental accessibility[12] and toxicity,[13] DuPont announced it will stop producing CFCs as substitute—hydrochlorofluorocarbon (HCFC, for short) and hydrofluorocarbons (HFC, for short)—became available in 1988 (Falkner 2005, pp. 108–111). Levy (1997, p. 63) argues, "with the support from the industry, the US position largely followed the stance of Dupont and the Alliance for Responsible CFC Policy". Namely, the US proposed an almost phase out of CFC consumption at the global level. The rational for this proposal is that, since without broad international compliance, foreign companies, especially European companies would be free to producing CFCs.[14] Consequently, the US chemical firms feared potential export markets for substitutes chemicals would fail.

The lesson from the corporate interest in ozone politics leads us to assume that the motivation to enhancing competitiveness through the business conflict influence state incentives for internationalization (Falkner 2008). De Sombre (2000) claims that corporate interest to enhance its competitiveness pushes for domestic internationalization. She claims:

> The international competitiveness effects of domestic environmental regulations are thus the necessary push, giving the regulated industries incentives to work for internationalization. In cases in which there is a cost to a given industry from the regulation, the industry should want its international competitors to bear the same cost. In the cases in which technology has been produced that responds to the regulation, industry also should want internationalization of the regulation in question to provide further markets for its products. In economic terms these are similar. In both cases there are competitive advantages for domestic industries if industries in other states are subject to similar regulations (De Sombre 2000, p. 45).

By drawing upon insights from the holistic constructivist approach and business conflict approach, this book proposes the new analytical framework of 'Agency with and beyond the State model' by combining the effects of domestic politics and international regulatory competition in order analyse the convergence of fuel economy regulations.

---

[11]The first mover's advantage or, Porter's hypothesis, claims that any companies would gain the advantage by occupying market segment.

[12]Alternative Fluorocarbons Environmental Acceptability Study in 1987.

[13]Program for Alternative Fluorocarbon Toxicology Testing in 1988.

[14]European companies were the biggest producer of CFC in 1986, that produced 48 % of the world share compare to the US which produced 28 %.

2.4 Automobile Industry in Global Climate Governance: 'Agency with and Beyond the States'

As mentioned, the 'Agency with and beyond the State model' is based on the 'Agency beyond the State model'. Agency here refers to an actor who possesses the ability to prescribe behaviour and to obtain the consent of the governed (Biermann et al. 2009, pp. 37–43). The key focus is placed on the need of the further investigations on whether nation-states can fulfil their core functions under the pressure of various environmental degradations, and to what extent non-state actors are filling new governance demands. More importantly, both approach enables us not only to look at the activities of non-state actors in earth system governance confined to "lobbying and advising national governments in the creation and implementation of rules", but also their roles as agency where they "substantially participate in/or set their own rules related to the interactions between humans and their natural environment" (Biermann et al. 2009, pp. 37–43).

However, what differentiates between the two approaches is while the latter tends to focus on private environmental governance (e.g. certification), the focus of the former is placed on how the industry that operates globally but very much engaged with their national governments, is operating a role as 'Agency' in international environmental politics.

The case of car industry fits well in the latter model, given that much of the climate policies relating to the automobile industry have been made at either a national (Japan and the US) or regional level (the EU), and making their approach one of 'Agency *with* and beyond the States' rather than 'Agency *beyond* the States'.

By using the Agency with and beyond the State model, this book contributes to bridge the enduring gap between the disciplines of IR and environmental studies by challenging the existing studies in this field. It adds new insight to the role of non-state actors in international environmental politics by showing how the automobile industry which operates globally but its strategy is strongly attached to national governments, can be Agency to bring about changes towards low-carbon society.

For this purpose, data used for analysis was collected through several research visits to relevant countries between 2007 and 2012, and includes primary documents with limited access, or those that have not been officially published. The main method of gathering data was semi-structured interviews conducted with relevant policy makers, directors and managers of automobile industries and its industry network, environmental NGOs and scientists and academics. Each interview lasted from one to three hours, with topics based on the fuel economy regulation of passenger cars, providing different perspectives on the development of actors, networks and institutions relating to the regulation. Informants were selected based on the following criteria: the person should have or at least have had definitive role in the development of fuel economy regulation. The careful selection of very central informants and generous time for each interview contributed to the quality. Furthermore, excerpts from each interview were submitted to the informants for review and approval. In addition, during the process of the interviews, several unpublished documents of great relevance for the study were obtained from the informants. Official government reports, media articles, policy papers by NGOs and research institutes, and other related materials were used to supplement the analysis.

References

Adler E (1997) Seizing the middle ground. Eur J Int Relat 3:319–364

Bang G et al (2005) Future U.S. climate policy: international re-engagement? Int Stud Perspect 6 (3):285–303

BASD (n.d.) What is BASD? Retrieved from http://basd.free.fr/about/index.html

Biermann F (2007) "Earth system governance" as a crosscutting theme of global change research. Glob Environ Change 17:326–337

Biermann F et al (2009) Earth system governance: people, places and the planet. Science and implementation plan of the earth system governance project. Bonn

Bäckstrand K (2008) Accountability of networked climate governance: the rise of transnational climate partnerships. Global Environ Politics 8(3):74–102

Bendell J (2000) Terms for endearment: business, NGOs and sustainable development. Greenleaf, Sheffield

Berger PL, Luckmann T (1967) The social construction of reality: a treatise in the sociology of knowledge. Anchor, New York

Bernstein S, Cashore B (2007) Can non-state global governance be legitimate? An analytical framework. Regul Governance 1(4):347–371

Börzel TA, Risse T (2005) Public-private partnerships. Effective and legitimate tools of transnational governance? In: Grande E, Pauly LW (eds) Complex sovereignty: on the reconstitution of political authority in the 21st century. University of Toronto Press, Toronto, pp 195–216

Bramble BJ, Porter G (1992) Non-governmental organizations and the making of US international environmental policy. In: Hurrell A, Kingsbury B (eds) The international politics of the environment: actors, interests, and institution. Clarendon Press, Oxford, pp 313–353

Carroll WK (2007) Hegemony and counter-hegemony in a global field. Stud Soc Justice 1(1):36–66

Cashore B (2002) Legitimacy and the privatization of environmental governance: how non state market-driven (NSMD) governance systems gain rule making authority. Governance 15(4): 503–529

CCX (n.d.) About CCX. Retrieved 2 Apr 2012, from https://www.theice.com/ccx.jhtml

Chan S, Pattberg P (2008) Private rule-making and the politics of accountability: analyzing global forest governance. Global Environ Politics 8(3):103–121

Christoff P, Ekersley R (2007) The Kyoto protocol and the Asia Pacific partnership on clean development and climate. In: Bonyhady T, Christoff P (eds) Climate law in Australia. Federation Press Sydney, Sydney, pp 32–45

Clapp J (1998) The privatization of global environmental governance: ISO 14000 and the developing world. Glob Gov 4:295–316

Clapp J (2003) Transnational Corporate Interests and Global Environmental Governance: Negotiating Rules for Agricultural Biotechnology and Chemicals. Environ Politics 12(4):1–23

Clapp J (2005) Global environmental governance for corporate responsibility and accountability. Global Environ Politics 5(3):23–34

Connelly J et al (2002) Politics and the environment: from theory to practice, 2nd edn. Routledge, London

De Sombre E (2000) Domestic sources of international environmental policy: industry, environmentalists, and US power. MIT Press, Cambridge
Dellas E et al (2011) Agency in earth system governance: refining a research agenda. Int Environ Agreements: Politics Law Econ 11(1):85–98
Desombre ER (2005) Fishing under flags of convenience: using market power to increase participation in international. Global Environ Politics 5(4):73–94
Doyle T, McEachern D (1998) Environment and politics. Routledge, London
European Commission (2011) How effective is the Forest Stewardship Council certification schemes. Brussels
Falkner R (2001) Business conflict and US Policy. In: Harris PG (ed) The environment, international relations, and U.S. foreign policy. Georgetown University Press, Washington, DC, pp 157–177
Falkner R (2003) Private environmental governance and international relations: exploring the links. Global Environ Politics 3(2):72–87
Falkner R (2005) The business of ozone layer protection: corporate power in regime evolution. In: Levy DL, Newell PJ (eds) The business of global environmental governance. MIT Press, Cambridge, pp 105–134
Falkner R (2008) Business power and conflict in international environmental politics. Palgrave McMillan, Basingstoke
Finnemore M (1996) National interests in international society. Cornell University Press, New York
Giddens A (1986) The constitution of society: outline of the theory of structuration. University of California Press, Oakland
Goldstein J (1996) International law and domestic institutions: reconciling North American "unfair trade" laws. Int Organ 50:541–564
Gourevitch P (1996) Squaring the circle: the domestic sources of international cooperation. Int Organ 50:349–373
Haas PM (1992) Introduction: epistemic communities and international policy coordination. Int Organ 46(1):1–35
Hall RB (1999) National collective identity: social constructs and international systems. Columbia University Press, New York
Hall PA, Soskice DW (2001) Varieties of capitalism: the institutional foundations of comparative advantage. Oxford University Press, Oxford
Hurrell A, Kingsbury B (eds) (1992) The international politics of the environment: actors, interests, and institution. Clarendon Press, Oxford
Jacobs M (1991) The green economy: environment, sustainable development and the politics of the future. Pluto, London
Jagers SC, Stripple J (2003) Climate governance beyond the state. Glob Gov: Rev Multilateralism Int Organ 9(3):385–399
Katzenstein P (1996) Cultural norms and national security: police and military in postwar Japan. Cornell University Press, Ithaca
Keidanren (1997) Keidanren voluntary action plan on the environment. Tokyo
Keohane RO (1988) International institutions: two approaches. Int Stud Quart 32:379–396
Keohane R, Axelrod R (1985) Achieving cooperation under anarchy: strategies and institutions. World Politics 38(1):226–254
Keohane R, Levy M (1996) Institutions for environmental aid: pitfalls and promise. MIT Press, Cambridge
Keohane R, Nye J (2000) Power and interdependence, 2nd edn. Harper Collins, New York
Krasner S (1983) International regimes. Cornell University Press, New York
Kratochwil F (1989) Rules, norms and decisions. Cambridge University Press, Cambridge
Lawrence P (2007) The Asia Pacific partnership on clean development and climate (AP6): a distraction to the Kyoto processor a viable alternative? Asia Pac J Environ Law 10(3/4): 183–209

Levy D (1997) Business and international environmental treaties: ozone depletion and climate change. Calif Manag Rev 39(3):54–71

Levy D (2005) Business and the evolution of the climate regime: the dynamics of corporate strategies. In: Levy DL, Newell PJ (eds) The business of global environmental governance. MIT Press, Cambridge

Levy D, Newell P (2002) Business strategy and international environmental governance: toward a neo-gramscian synthesis. Global Environ Politics 2(4):84–101

Levy D, Newell P (eds) (2005) The business of global environmental governance. MIT Press, Cambridge

Lipschutz RD (2004) Limitations of empire. Global Environ Politics 2004(May):20–23

McGee J, Taplin R (2006) The Asia-Pacific partnership on clean development and climate: a complement or competitor to the Kyoto protocol? Glob Change Peace Secur 18(3):173–192

Mikler J (2006) Varieties of capitalism: national institutional explanations of environmental product developments in the car industry. The University of Sydney, Sydney

Mikler J (2007) Varieties of capitalism and the auto industry's environmental initiatives: institutional explanations for firms' motivations. Bus Politics 9(1):1–38

Mikler J (2008) Sharing sovereignty for global regulation: the cases of fuel economy and online gambling. Regul Governance 2(4):383–404

Mikler J (2009) Greening the car industry: varieties of capitalism and climate change. Edward Elgar, Cheltenham

Mikler J (2010) Business as usual or apocalypse now? Regulating the car industry's carbon emissions. In: Third Biennial conference of the european consortium on political research standing group on regulatory governance. Dublin

Newell P (2001a) Environmental NGOs, TNCs, and the question of governance. In: Stevis D, Assetto VJ (eds) The international political economy of the environment. Lynne Reinner, London, pp 85–107

Newell P (2001b) Managing multinationals: the governance of investment for the environment. J Int Dev 13:907–919

Newell P (2005) Business and international environmental governance: the state of the art. In: Levy DL, Newell P (eds) The business of global environmental governance. MIT Press, Cambridge, pp 21–45

Newell P, Paterson M (1998) A climate for business: global warming, the state and capital. Rev Int Polit Econ 5(4):679–703

Onuf N (1989) World of our making. University of South Carolina Press, Columbia

Paarlberg RL (1996) A domestic dispute: Clinton, Congress, and international environmental policy. Environment 38(8):16–20

Pattberg P (2005) What role for private rule-making in global environmental governance? analysing the Forest Stewardship Council (FSC). Int Environ Agreements: Politics Law Econ 5 (2):175–189. doi:10.1007/s10784-005-0951-y

Pattberg P (2007) Private Institutions and Global Governance. The New Politics of Environmental Sustainability. Cheltenham: Edward Elgar

Pattberg P, Stripple J (2008) Beyond the public and private divide: remapping transnational climate governance in the 21st century. Int Environ Agreements: Politics Law Econ 8(4): 367–388

Pearse R (2010) Making a market? Contestation and climate change. J Aust Polit Econ 66:166–198

Porter G, Brown JW (1996) Glob Environ Politics. Westview Press, Boulder

Pratt N (2004) Bringing politics back in: examining the link between globalisation and democratization. Rev Int Polit Econ 11(2):311–336

Raustiala K (1997) Domestic institutions and international regulatory cooperation. World Politics 49(4):482–509

Reus-Smit C (2005) Constructivism. In: Burchill S, Linklater A, Devetak R, Donnelly J, Paterson M, Reus-Smit C, True J (eds) Theories of international relations, 3rd edn. Palgrave McMillan, Basingstoke, pp 188–211

Rosenau JN (1995) Governance, order, and change in world politics. Governance without government: order and change in world politics. Cambridge University Press, Cambridge
Rowlands I (1995) The politics of global atmospheric change. Manchster University Press, Manchester
Rowlands I (2001) Transnational corporations and global environmental politics. In: Wallace W, Josselin D (eds) Non-state actors in world politics. Palgrave McMillan, Basingstoke, pp 133–149
Ruggie J (1998) Constructing the world polity. Routledge, London
Schreurs MA, Economy E (1997) The internationalization of environmental protection. Cambridge University Press, Cambridge
Smith S (1999) Social constructivisms and European studies: a reflectivist critique. J Eur Public Policy 6(4):682–691
Source Watch (n.d.) Global climate coalition. Retrieved 29 Mar 2012, from http://www.sourcewatch.org/index.php?title=Global_Climate_Coalition
Sussman G (2004) The USA and global environmental policy: domestic constraints on effective leadership. Int Polit Sci Rev 25(4):349–369
Underdal A, Hanf K (2000) International environmental agreements and domestic politics: the case of acid rain. Ashgate, Aldershot
Usui M (2002) The role of private business in international environmental governance. Working Paper IAS–CGP Project on Intl Environ Governance, pp 1–31
Van Asselt H (2007) From UNI-ty to diversity? The UNFCCC, the Asia-Pacific partnership, and the future of international law on climate change. Carbon Climate Law 1(1):17–28
Vogel D (2003) The hare and the tortoise revisited: the new politics of consumer and environmental regulation in Europe. Br J Polit Sci 33(4):557–580
Waltz K (1959) Man, the state and war. Columbia University Press, New York
Wapner P (1995) Politics beyond the state: environmental activism and world civic politics. World Politics 47(3):311–340
WBCSD (n.d.) Overview. Retrieved 5 Mar 2013, from http://www.wbcsd.org/about.aspx
Weaver K, Rockman B (1993) Do institutions matter? Brookings Institution, Washington DC
Webb K (2002) Voluntary Codes: Private Governance, The Public Interest and Innovation. Otawa: Carlton University
Wendt A (1987) The agent-structure problem in international relations theory. Int Organ 41(3): 335–370
Wendt A (1992) Anarchy is what states make of it: the social construction of power politics. International Organization 46(2):391–425
Wendt A (1999) Social theory of international politics. Cambridge University Press, Cambridge
WWF (n.d.) Making fishing sustainable
Young OR (1999) Governance in world affairs. Cornell University Press, Ithaca

Chapter 3
Construction of European Fuel Economy Regulations for Passenger Cars

Abstract This chapter looks at how Europe's fuel economy regulations for passenger cars have been constructed. As demonstrated in Introduction, Europe currently has the highest fuel economy standards in the world. How and why has the EU introduced these standards, even before Japan and the United States? What factors influenced these regulations, and which actors were instrumental in the decision-making process? This chapter argues that the European fuel economy regulations have been progressed by conflicts over business interests between Germany and France, both of which are motivated to increase the competitiveness of its automobile industry within the European market. Backed up by the EU's dynamic decision-making process, business conflicts between Germany and France in turn led to the EU to commit the strictest fuel economy regulations in the world that triggered the regulatory convergence. The primary focus is placed on supranational decision-making at the EU level. This is because fuel economy regulations are formulated as EU law, which supersedes the domestic environmental regulations of the member states. The positions and interests of the member states will be also mentioned in order to understand who influenced the process of establishing the EU fuel economy regulations, as well as how and why.

Keywords Europe · Car fuel economy regulations · Competitiveness of European automobile industry · Conflict between Germany and France · Supranational decision-making

3.1 Introduction: EU as a Normative Power

This chapter looks at how Europe's fuel economy regulations for passenger cars have been constructed. As demonstrated in Introduction, Europe currently has the highest fuel economy standards in the world. How and why has the EU introduced these standards, even before Japan and the United States? What factors influenced these regulations, and which actors were instrumental in the decision-making process?

M. Iguchi, *Divergence and Convergence of Automobile Fuel Economy Regulations*,
DOI 10.1007/978-3-319-17500-3_3

To answer this question, it would be wise to first look at a growing body of literature on the EU's role as a global actor (Delreux 2012) and its 'normative power' in international environmental politics (Manners 2002). This literature looks at the EU's central role in creating and leading climate politics through its ideas and values (Vogler and Bretherton 2006), and broadly the sustainable development at the United Nations (Lightfoot and Burchell 2005). Kelemen (2010) offers an insightful explanation for the EU's leadership on global environmental regulation from two viewpoints. First, the growing environmental interests in Europe beginning in the late 1980s, coupled with dynamic EU policy-making, led the EU to commit to ambitious environmental policies. Second, the EU's international competitive interests led the EU to support international agreements that would pressure other countries to adopt similar environmental regulations, which, in turn, spread the EU environmental norms to other regions and served to legitimize the EU's rules prior to any world trade agreements.

In fact, the EU's environmental policy emerged out of the concerns among member states that diverse environmental standards could result in trade barriers and competitive issues in the common market (Johnson and Corcelle 1989). In particular, differing national environmental standards on vehicle emissions and the lead content of petrol "posed [a] formidable obstacle to the free trade of these products within the Economic Community (EC) … Against this background, EU environmental policy was primarily a policy flanking the Common Market … They were motivated instead by competition policy, or to be more precise, the realization of the Common Market by harmonizing national legal and administrative regulations" (Knill and Lieffernik 2007, p. 14).

The EU's environmental policies were developed after the UN Conference on the Human Environment in 1972, which set a milestone of global environmental governance.[1] In the following year, the first European Environmental Action Programme was established; it emphasized that economic and environmental issues have to be mutually constitutive. Since then, European environmental policy has developed significantly as the EU member states have become more integrated. Table 3.1 shows how the Maastricht, Amsterdam, Nice, and Lisbon treaties enhanced the EU's environmental policy over the following years in terms of integrating environmental principles into its decision-making process. As the integration deepened, qualified majority voting (the opposite to unanimous voting) and co-decision making procedures become standardized. This enabled swift environmental policy-making at the EU level and strengthened the role of the European Parliament, which is the only EU institution for which members are directly elected. This, in turn, created a space for the environmental pressure groups to push for higher environmental standards, including stringent fuel economy regulation.

[1]The UN Conference on the Human Environment (UNCHE) raised international awareness on human interaction with the environment. The conference was attended by 113 countries, 19 inter-governmental agencies and more than 400 inter-governmental and non-governmental organizations.

Table 3.1 Advancement of the EU environmental policy

Year signed	Year in force	Treaty	Changings affective environmental policy
1957	1958	Rome	No mention of environment
1986	1987	Single European act	Environmental title added
			Article on environmental policy integration added qualified majority voting for the internal market
1992	1993	Maastricht	'Sustainable growth respecting the Environment' becomes one of the tasks of the community (article 2)
			Environment title strengthened to include mention of 'precautionary principle'
			Integration of article (article 120r) was reinforced
			The number of policy areas where the council could adopt environmental legislation using QMV was extended
			Co-decision strengthened the role of European parliament in developing environmental policy
1997	1999	Amsterdam	Article 2 strengthened so that 'sustainable development of economic activities' made an explicit objective of the EU
			Integration article given more prominence (article 6)
			Co-decision becomes the normal process for agreeing environmental policy
2001	2003	Niece	QMV changed to establish a double majority of member states and votes cast
2007	2009	Lisbon	Environment Title (174–176, TEC) substantially unchanged but numbering changed (now articles 191–193, TFEU)
			Integration article now article 11
			Article 2 strengthened so that the EU shall work for the 'sustainable development of Europe' and the 'sustainable development of the earth'

Source directly quoted from Jordan and Adelle (2012, p. 4)

This chapter argues that the European fuel economy regulations have been progressed by conflicts over business interests between Germany and France, both of which are motivated to increase the competitiveness of its automobile industry within the European market. Backed up by the EU's dynamic decision-making process, business conflicts between Germany and France in turn led to the EU to commit the strictest fuel economy regulations in the world that triggered the regulatory convergence.

There is no doubt that the automobile industry plays a vital role in Europe's economy. The industry creates huge amounts of employment, with 2.3 million direct jobs and 10 million indirect jobs, and car taxes contribute to 4.1 % of EU's gross domestic product (ACEA 2009). Europe is the home to major automobile companies, including Mercedes-Bentz (founded in 1886), Volkswagen (founded in 1938),

Peugeot (founded in 1882), Citroen (founded in 1919), and Fiat (1899), and these companies have formed the basis of the economies in their countries. In addition, three of these automotive companies, Volkswagen (Germany), PSA Peugeot Citroen (France), and Fiat (Italy), have large shares in the global automobile market. Despite the fact that the EU has successfully reduced its CO_2 emissions,[2] the emission from the road transport sector is on the rise. In 2009, the road transport sector in the EU accounted for about 24 % of all CO_2 emissions (855.6 million tons of CO_2) (IEA 2009).[3] Therefore, mitigating car CO_2 emissions in the EU is a high-stakes issue. In order to analyse the development of the European fuel economy regulations, following sections are divided into three periods: the formative years during the early 1990s; the appearance of the 120 g/km target in the late 1990s; the EU industry voluntary agreement in 1998–2006; the critical juncture in 2007; and legalizing the 120 g/km target and beyond from 2007 to 2009.

The next section looks at how European fuel economy regulations[4] have been constructed. The primary focus is placed on supranational decision-making at the EU level. This is because fuel economy regulations are formulated as EU law, which supersedes the domestic environmental regulations of the member states. The positions and interests of the member states will be also mentioned in order to understand who influenced the process of establishing the EU fuel economy regulations, as well as how and why.

3.2 The Formative Years: The Early 1990s

In the early 1990s, concerned with increasing car CO_2 emissions, Europe started to consider introducing fuel economy regulations. In Europe, fuel economy for cars was regulated at the European level, not at the member state level. This is because as the integration of the European member states deepened, free movement of people, goods, and the capital increased. Because of these changes, regulation at the member state level would be ineffective; hence, regulations at the supranational level were needed to limit growing CO_2 emissions from cars. The very first proposal to limit CO_2 emissions from cars was presented by the European Commission in response to

---

[2]In 1990, the total CO_2 emission from fuel combustion in EU was 4,051 million tonnes of CO_2, and the number has decreased to 3,576.8 million tonnes of CO_2 in 2009 (IEA 2009, p. 46).

[3]Transport sector accounts for 912.9 million tons of CO_2. Emissions from other sector are as follows: 1,337 million tons of CO_2 from electricity and heat production; 181.1 million tons of CO_2 from other energy industry own use; 891.9 million tons of CO_2 from manufacturing industries and construction; and 683.5 million tons of CO_2 from other sectors including residential sector.

[4]European fuel economy regulation was first expressed by Litres per 100 km (l/100 km) until 1994, but later expressed by CO_2 emissions per kilometres (g/km) since 1995 onwards. Therefore, it is more accurate to express 'European CO_2 emission reductions from new cars', but for the convenience, this book treats g/km target as 'fuel economy regulation', since the target is comparable with Japan and the US.

a request by the European Council to limit CO_2 emissions from the road transport sector. The Commission proposal was made as follows in 1991:

> [w]ith and average of 2.2 tons of carbon per head, the Community represents 13 % of global CO_2 emissions, compared to 23 % for the U.S., 5 % for Japan and 25 % for Easter Europe and the USSR. Four main sectors in the Community are responsible for these emissions: power generation (31 %), transport (26 %), Industry (20 %), and residential/ commercial (20 %). During the period 1970-1985 emissions almost stabilised. During the period 1986-1990, however, this positive tendency has been reversed and emissions have grown by 4 %...For the period 1990-2000, CO_2 emissions are likely to continue to grow by another 11 %...Transport is currently the source of around 25 % of the Community's CO_2 emissions. This share is liable to increase in the future, mainly as a consequence of the expected further growth in the volume of road traffic. Because road traffic also entails other considerable external costs (acid emissions, congestions etc.), structural policies are urgently needed at the Community level and in the Member States to encourage more environmental rational approach towards mobility (European Commission 1991).

What is apparent is that the origin of the European fuel economy regulations was built on the concerns for the growing CO_2 emissions from the road transport sector and the need of to apply the best available technology to reduce these emissions and increase the fuel efficiency of cars. Along with the concerns of growing CO_2 emissions, another purpose of introducing the regulation was enhancing the industry competitiveness. According to a Commission strategy paper in 1991, "an ambitious programme to improve the efficiency will increase energy security, improve energy efficiency of the transportation system, limit energy related air emissions other than CO_2 and can strengthen industrial competitiveness" (European Commission 1991, p. 4). Therefore, the European fuel economy regulations were introduced in response to the growing CO_2 emissions from cars as well as to enhance industry competitiveness by improving car fuel economy.

In order to implement this plan, the Commission was assigned by a directive on vehicle emissions to propose legislation to the Council of Environmental Ministers by the end of 1992. To this end, the Commission consulted with the Motor Vehicle Emission Group (MVEG), an advisory body comprised of officials from the member states and motor industry representatives (The ENDS Report 1992a). The MVEG met approximately every two months from its first meeting on 14th January 1985 until 1996 and consisted of 40–50 members. German participants were among the largest group in the MVEG, with up to eight officials from three different ministries, including the Ministry of Environment, Nature Conservation, and Nuclear Safety; the Ministry of Economics and Technology; and the Federal Environmental Agency (Wurze 2002, p. 140). Therefore, we can assume that Germany had the strongest voice in the group from the very beginning, which acted both the dragger and the pusher for the remainder of the discussions on the European fuel economy regulations.

Within the MVEG discussions, at least four major options were submitted from France, Germany, the United Kingdom (UK), and Italy to control CO_2 emissions from the road transport sector. France, which already produced relatively fuel-efficient cars, emphasized a need for setting European-level regulatory standards in terms of 'grams of CO_2 per kilometres (g/km)'. France also argued that those

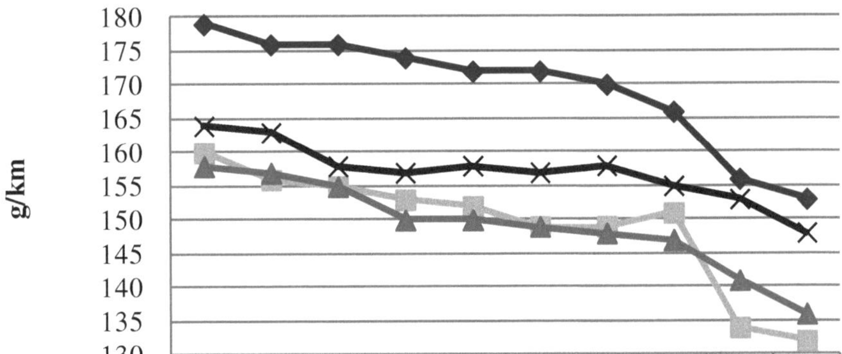

	2001	2002	2003	2004	2005	2006	2007	2008	2009	2010
Germany	179	176	176	174	172	172	170	166	156	153
France	160	156	155	153	152	149	149	151	134	132
Italy	158	157	155	150	150	149	148	147	141	136
EU-12	164	163	158	157	158	157	158	155	153	148

Fig. 3.1 A comparison of actual fuel efficiency of passenger automobiles between European countries, 2001–2010. *Source* created by author based on ICCT (2011)

companies that achieved the target could get financial benefits while those that exceeded the target would pay a fine. Germany, in contrast, produced heavy-weight cars and was in favour of weight-based fuel economy regulations,[5] insisting that CO_2 emission targets should be based on engine capacity or vehicle weight. The British government proposed a system of tradable emission credits, in which manufacturers that met a specified fuel efficiency standard would sell their credits to those whose products did not. Finally, Italy, which produced small cars, favoured a tax incentive system based on CO_2 emissions, where no tax would be payable on low-carbon vehicles (The ENDS Report 1992b). What is clear from these proposals is that each country had very different ideas of what the European-level regulations should look like. France and Italy, which already produced fuel-efficient cars, argued for regulatory incentives that fined manufacturers that did not achieve the satisfactory standards. Germany and the UK, which produced heavy luxury cars, supported flexibility measures, such as vehicle weight–based CO_2 regulations as well as a tradable emission credits in case manufactures could not achieve the target. Figure 3.1 is a comparison of the CO_2 emission from passenger cars among European countries from 2001 to 2010. It is clear that while France and Italy had already been significantly improving their fuel economy standards, Germany has been the most stagnant in this regard, well below the average standard within the EU.

[5]Under the weight-based fuel economy regulations, fuel economy regulation is differentiated according to the weight of vehicles, as long as the mean of all vehicles sold in Europe satisfied the desired target. This means that manufactures produce heavy cars would have to achieve less fuel economy efficiency that manufactures that produce lighter vehicles.

As a result of these divergent voices, MVEG discussion deadlocked, which made it difficult for the Commission to formulate the legislation that was originally expected to be completed by the end of 1992. In response, in 1991, the European car manufacturers established a lobbying industry network, the European Automobile Manufacturers' Association (ACEA) as a successor to the Comité des Constructeurs du Marché Commun manufacturers committee. The ACEA committed itself to a voluntary emission reduction target of **10 % between 1993 and 2005** (Keay-Bright 2000, p. 19). Therefore, the first European car climate policy took the form of industry self-regulation,[6] which was the very first attempt by industry to weaken the initial target of fuel economy regulation.

This industry proposal was not accepted by the MVEG. Instead, in November 1992, right after the UNFCCC was signed, a subgroup of MVEG decided to set a stringent fuel economy target of 40 % improvement with a system of charges added to the price of new cars to be implemented from 1995 to 2005. In this system, the baseline for the charges was set at zero for cars emitting no more than 160 g/km that were produced in 1995 and gradually reduced to 5 g/km annually until 2005 (an equivalent of 110 g/km by 2005).[7] Any new cars that failed to meet these targets would be charged, although this rate was up to the member states to decide (The ENDS Report 1992a). What we can observe from this conclusion is that the French proposals were well reflected in the proposal—namely, setting European-level regulatory standards expressed in terms of 'g/km' and fining those who exceeded the target. This French preference remained the case for the remainder of the regulatory process and thus became a foundation of Europe's CO_2 reduction–driven nature of the fuel economy regulation for cars.

Still, this conclusion of a 110 g/km by 2005 target proposed by the MVEG did not gain much support and gradually disappeared without a trace because it was faced with major opposition from financial ministers from each country.[8] Consequently, with the absence of an agreement between states and the automobile industry, the original legislative deadline assigned to the Europe Commission was missed.

---

[6]In existed literatures, industry self-regulation occurs when corporations design and enforce the rules themselves. These rules are generally adopted voluntarily. According to Haufler (2001, pp.8–9), there are two models of industry self-regulation. First model is to create a technical standard that specify the physical qualities required for sale and use of industrial or commercial products and services, for the sake of market promotion. Good example of this model is International Organization for Standarzation (ISO). The second model is based on social or political demands from outside the business community, namely, safeguard prior to government regulation. The context of the ACEA self-regulation in setting its fuel economy regulation falls into the second model.

[7]This would be equal to 40 % improvement of fuel economy of cars (The ENDS Report 1992b).

[8]They opposed on using fiscal system for environmental ends at the MVEG meeting, held on 9th December 1992 (The ENDS Report 1992a, 1996).

3.3 The Appearance of the 120 g/km Target: The Late 1990s

A major breakthrough took place in the Environment Council held in October 1994. Prior to the Council meeting, two international events are worth noting as a background of the proposal. Firstly, the UNFCCC was born out of the 1992 UN Conference on the Environment and Development in Rio de Janeiro with the aim to stabilize greenhouse gas concentrations in the atmosphere. Secondly, the 1993 Maastricht Treaty incorporated 'sustainable growth respecting the environment' and 'precautionary principle' into the tasks of the Community. It was these international events that opened the window for further European climate policy development.

In October 1994, Germany proposed to increase the fuel economy of new cars to an average of 5 L/100 km (120 g/km) for gasoline cars and 4.5 L/100 km for diesel cars by 2005 in the 1791st Council meeting on environment under the German presidency (Council of the European Union General Secretariat 1994; The ENDS Report 1994). Later, the 120 g/km target formally appeared in a communication from the Commission to the Council and the European Parliament in December 1995. It states:

> [t]he Environment Council in December 1994 more specifically requested the Commission to look into the possibility of substantially lowering the fuel consumption of newly registered cars by 2005. In this context, an average fuel consumption of 5L/100 km for petrol cars and of 4.5L/100 km for Diesel cars (equivalent to 120 g/km) has been mentioned by twelve Member States and the European Parliament as a target (Commission of the European Communities 1995, p. 4).

It is also significant that this target was agreed upon by the German Environment Minister, Angela Merkel (in office from November 1994 to October 1998), who later strongly opposed legalizing the 120 g/km target in 2007 as Prime Minister of Germany. The question is, then, why Germany, among other automobile manufacturing member states, proposed stringent fuel economy regulations. One answer can be found in the industry position of that time - the industry was not strongly opposed to adopting the proposed target. Rather, the German auto industry had been proactively seeking to improve vehicle emission standards since the late 1970s. Such exertions were influenced by the advanced US car air pollution regulations, especially California's stringent standards, as this was where the German auto industry was selling half of their products in the United States. Later, the German car industry supported the European Community in adopting the American air pollution standards (Volgel 1997, p. 562). In fact, Germany was leading the emission-reduction technology development in Europe, especially regarding the emissions of sulphur dioxide and nitrogen dioxide (Weidner 1995, p. 46). So does the fuel efficiency of cars. In 1978, the German auto industry actually pledged to improve fuel efficiency by 15 % from 1978 to 1985. After this pledge was satisfied, it extended its pledge in 1995, committing to a 25 % reduction in the average fuel consumptions of the cars that were sold in Germany between 1990 and 2005 (Commission of the European Communities 1995, p. 13). Therefore, the reason behind Germany's proposal can be explained as a desire to enhance its market competitiveness. According to a campaigner of the Greenpeace Europe:

> Up to those days there were probably an optimism that German car manufactures could achieve this target by technological innovation. Germans also made a positive role in the discussion about pollutant standards which was considered that the companies to gain technology leadership and market competitiveness in the international field. Therefore the companies was not opposed the standards in the early 1990s when this things was discussed.[9]

Despite Germany's willingness to improve stringent fuel economy regulations, once the 120 g/km target was mentioned in the communication from the Commission to the Council and the European Parliament in December 1995, ACEA started to lobby against a timetable for the target in order to postpone the original target year of 2005 (Greenpeace 2008). As a result, the target timetable was extended 5 years past the original year. This can be seen in various documents released by each European institution (Commission of the European Communities 1995; Council of the European Union General Secretariat 1996; EESC 1998). These documents include the communication released by the Commission in 1995, that on the one hand claimed that 'significant progress' could be made in order to achieve the 5 L/100 km target for petrol cars; but it also noted that achieving this target by 2005 was 'rather ambitious', and thus urged the ACEA to reduce average CO_2 from new cars by 25 % between 1990 and 2005 and to meet 120 g/km **by 2010** (Commission of the European Communities 1995). Another evidence of industry influence over the timetable for reaching the target can be seen in the texts adopted in the Council. In the press release of the 1939th Council meeting, while it affirmed the medium target of 120 g/km by 2005, it stated "[s]hould it appear that it is not possible fully to achieve the objective of 2005, the phasing could be extended, **but in no case beyond 2010**" (Council of the European Union General Secretariat 1996). Furthermore, a text released by the European Economic and Social Committee (EESC) added, **"in no case beyond 2010"** to the 2005 target year (EESC 1998). What is apparent from these texts is that industry had successfully lobbied its preference to delay the target year.

The question then arises of why the ACEA attempted to delay the target timetable, despite Germany proposed 120 g/km target in 1995? One of the answers can be found in the conflict between the German and French automobile manufacturers —the two countries that were driving the European integration—over the best approach to fuel economy regulation. Needless to say, Germany and France were—and still are—the major automobile manufacturing countries within Europe as indicated in Table 3.2, and they had the highest shares of passenger cars in the European market: Germany shares about 20–25 %, followed by followed by France and Italy that shares around 15 % respectively.

In the discussion over the supranational fuel economy regulations, while German manufacturers pushed for a percentage reduction target of 75 % against 1990 levels by 2005 and the introduction of a weight-based system, French and Italian

[9]Based on interview with Ms. Franziska Achterberg, Greenpeace EU transport campaigner. Interviewed at Brussels, Belgium (21st May 2012).

Table 3.2 Share of passenger cars sold in Europe, by member states 2000–2010

Year/Country	Germany (%)	France (%)	Italy (%)
2001	22	15	16
2002	22	14	16
2003	22	13	15
2004	21	13	15
2005	22	14	15
2006	22	13	15
2007	20	13	16
2008	22	14	15
2009	27	16	15
2010	22	17	15

Source created by author based on ICCT (2011)

manufacturers insisted on regulation based on absolute figures expressed by g/km with incentives for those who achieved the target, such as financial benefits, while those who exceed the target would pay a fine (The ENDS Report 1997). The German preference, namely the introduction of a weight-based system, suggested that fuel economy regulation could be differentiated according to the weight of vehicles, as long as the mean of all vehicles sold in Europe satisfied the 120 g/km target. This meant that the French and Italian manufacturers, who produced lighter cars, would have to carry a heavier burden for improving fuel efficiency than the German manufacturers. Therefore, it was only natural to assume that the French and Italian automobile industry wanted an absolute target of 120 g/km for every manufacturer, regardless of the weight of the vehicle. Furthermore, they preferred a regulatory incentive that would give financial benefits to achievers and fines to underachievers. This, in turn, meant that the German manufacturers would have to take on considerable burdens to satisfy the target. Hence, we can assume that more time was needed to coordinate the best strategy within the industry network as a result of these clashes of interests. The remainder of the sections demonstrates that the clashes of interests between Germany and France (and, to an extent, the Italian) were critical to the construct of the European fuel economy regulations. The position of the ACEA, in turn, seems to be adopted as the lowest common denominator between the German and French manufacturers. The next section focuses on the voluntary agreement between the European Commission and the ACEA, which was concluded as a result of negotiations.

3.4 EU-Industry Voluntary Target: 1998–2006

As an alternative to the 120 g/km target, ACEA offered in 1997 to improve average CO_2 emissions to '155 g/km by 2005', which was equal to a nine per cent improvement against the average CO_2 emissions of 171 g/km at that time.

Nonetheless, the ACEA's '155 g/km by 2005' target was based on calculation under the old test cycle; under the new test cycle, this target converts to 167 g/km, and it was therefore far from the 120 g/km target (The ENDS Report 1997, 1998a). This implies that the European auto manufacturers actually attempted to undermine the 120 g/km target, and hence refused by the 2062nd Council meeting. In the meeting, the Council reaffirmed the importance of the 120 g/km target by 2005 or 2010 at the latest and noted that "the offer made by the motor industry in its negotiation with the European Commission on a voluntary environmental agreement was quite inadequate" and insisted that it is necessary that "the motor industry take action to ensure a satisfactory outcome of the negotiation" (Council of the European Union General Secretariat 1997).

In reaction, in March 1998, the ACEA outlined a proposal, including a voluntary target of 140 g/km by 2008 on the newer test cycle and having some 120 g/km models available by 2000 (The ENDS Report Report 1998b). Facing the industry opposition to pursue '120 g/km by 2005' target, the Council compromised that this proposal could be the basis for further negotiation leading to an agreement at its 2106th Council meeting (Council of the European Union General Secretariat 1998). As a result, a *voluntary* target between the ACEA and the European Commission (1999/125/EC) was agreed upon in 1998 with a target of 140 g/km by 2008. Most important, the timetable for 120 g/km was modified to 2012 from the original year of 2005, in accordance of the first commitment period of the Kyoto Protocol in mind (The ENDS Report 1998c; Commission Staff of the European Communities 1998; EESC 1998).

Consequently, European automobile industry acted as a laagered to delay the timetable for 120 g/km target as well as to weaken the target itself. Therefore, a voluntary agreement of '140 g/km by 2008' target was a result of a 'compromise' or a 'political exchange' between the ACEA and the European Commission (Usui 2007): On the one hand, by offering to reduce CO_2 emissions to 140 g/km by 2008 —in the original proposal by the Commission, the timetable for 120 g/km was 2005, or 2010 at the latest—the ACEA had successfully avoided the mandatory CO_2 emission reduction under the EU law, which was what the European Commission had been insisting upon; On the other hand, the European Commission wanted to implement a CO_2 reduction target for passenger cars, and hence accepted a voluntary agreement.

The next section explains why Europe moved to introduce legally binding fuel economy regulations despite the industry having been successfully avoiding such regulations.

3.5 Critical Juncture: 2007

Prior to the final year of the ACEA's voluntary target, the European Commission proposed to legalize the 120 g/km target on 19th December 2007. Its proposal was to reduce the average emissions of CO_2 from passenger cars in the EU from around

160 to 130 g/km by 2012 and to achieve 120 g/km using an integrated approach (European Commission 2007a). Apart from the focus on reducing CO_2 emissions from passenger cars, it also aimed to safeguard the competitiveness of the car industry by stimulating the development and deployment of cutting-edge automotive technologies (European Commission 2007c).

There were several factors that pushed the Europe's mandatory 120 g/km target, including the peoples' interests in climate change, concerns regarding energy security, and desires to reduce dependence on foreign oil.[10] However, the single most important reason why the Commission had announced 120 g/km target again was that the industry was way below the voluntary agreements standards and they therefore needed a regulatory measurement.[11] In other words, the car manufacturers had made a commitment to the voluntary agreement but had not respected it. The roles of the environmental NGOs were particularly important for constructing a legislative process for the 120 g/km target. The role of the NGOs was to be a 'watch-dog', speaking up about what mattered, and exposing the corporate lobby, as well reminding policy makers about the issues at stake by contacting them directly, through the media, and through public hearings.[12] Most importantly, T&E, one of the major NGOs that specialized in CO_2 and transport, started to publicize the performance of each manufacturer to show that they were not respecting or striving toward the proposed target. The ACEA did not want this information to be made available to the public, so the agreements with the Commission were clear that no manufacturers' specific data should be showed to the public. Instead, there was a consensus that the data could only be shown as from the ACEA, the Japan Automobile Manufactures' Associations (JAMA), and the Korean Automobile Manufactures' Association (KAMA). This issue become a 'public item' and drew the attentions of the public interests and advanced the discussions of the legalization of the 120 g/km target.

It became apparent that despite the ACEA's 140 g/km voluntary target, the average fuel economy of the European car industry in 2006 was only 160 g/km (T&E 2007a). This negative shift in the reputation of the automobile industry helped to create more space for arguments made by civil society in support of stringent fuel economy standards under a regulatory approach. Therefore, European civil society was critical in forcing the juncture of the European fuel economy regulations by revealing that the voluntary agreement had clearly failed. This, in turn, reconstructed the public debate on the fuel economy regulation and forced the European Commission to legalize the 120 g/km target.

Along with the pressure from domestic NGOs to push for a mandatory 120 g/km target, pressure from international commitments was also important for constructing

[10]Based on interview with Dr. Peter Mock, Europe Lead, International Council on Clean Transportation (ICCT). Phone interview (18th May 2012).

[11]Based on interview with Mr. Greg Archer, Programme Manager, Transport and Environment (T&E). Interviewed at Brussels, Belgium (22nd May 2012); and interview with Dr. Peter Mock, ICCT.

[12]Based on interview with Ms. Franziska Achterberg, Greenpeace EU.

the European fuel economy regulations. The EU committed to reducing CO_2 emission by eight per cent (based on the 1990 level) between 2008 and 2012 under the Kyoto Protocol. Given that the road transport sector was one of the major sources of CO_2 emissions within the EU, emissions reduction was critical to achieve its commitment. In the Commission strategy adopted in 2007, it emphasises:

> [r]oad transport is the second largest GHG emitting sector in the EU. It remains one of the few sectors whose emissions keep rising, thereby jeopardising the progress made by other sectors. This makes it harder for the EU to meet its Kyoto commitments and has negative repercussions on the competitiveness of certain sectors (e.g. energy intensive industries) which are also sensitive to international competition than domestic activities such as road transport (European Commission 2007b).

Furthermore, legalizing the 120 g/km target was important for EU's strategy for establishing international climate policies beyond the Kyoto Protocol. When the 120 g/km legislation was proposed by the European Commission, there was a wide range of other climate policies that were proposed at the same time for the Copenhagen Conference, such as the EU's energy and climate package that set 20 % reduction of GHG emissions against 1990 level by 2020. Although the car fuel economy regulation was not formally part of the package, but was adopted at the same time, there was a big political push to put in place a wide range of stringent climate policies.[13] Therefore, both pressures from the domestic NGOs and EU's international commitment encouraged the EU to legalize the 120 g/km target. The next section focuses on the construction of EU's fuel economy regulation targeted for 2015 and beyond.

3.6 Target for 2015 and Beyond

Twelve years after the Commission first proposed the introduction of the '120 g/km by 2005' target, the Commission once again proposed the same target in 2007. The impact of the regulation on the European automobile industry was valid, since the major European auto manufacturers, such as Volkswagen, PSA, and Fiat, produced and sold most of their vehicles within Europe. Table 3.3 describes the detailed share of motor vehicle production and net sales by these manufacturers.

Although the original year for introducing the target had been delayed due to industry lobbying, the numerical target itself had survived. This suggests that once a number has been placed on the supranational negotiation table in the European political system, it is hard to remove. Such dynamics are born out of codecision procedure between the European Commission, the Parliament, and the Council.

[13]Based on an interview with a Policy Officer, Transport and Ozone Unit, Climate Action Directorate General, European Commission. Interview at Brussels, Belgium (15th May 2012).

Table 3.3 Motor vehicle production and net sales by major European automobile industry by region, 2007–2010

Year/Company		Volkswagen (Germany)	PSA (France)	Fiat (Italy)
2007	Production	**EU: 4,100,010 (65 %)**	**EU: 2,742,917 (79 %)**	**EU: 1,683,324 (63 %)**
		America: 1,199,902 (19 %)	America: 227,128 (7 %)	America: 727,399 (27 %)
		Asia: 855,427 (14 %)	Asia: 466,621 (13 %)	Asia: 78,069 (3 %)
	Sales	**EU: 60.4 %**	**EU: 84.9 %**	**EU: 67.1 %**
		North America: 8.2 %	Latin America: 5.7 %	North America: 9.9 %
		Asia: 17.4 %	Rest of the world: 9.2 %	Rest of the world: 22.8 %
2008	Production	**EU: 4,124,820 (64 %)**	**EU: 2,477,812 (75 %)**	**EU: 1,500,643 (59 %)**
		America: 1,286,340 (20 %)	America: 266,110 (8 %)	America: 774,176 (31 %)
		Asia: 871,795 (14 %)	Asia: 506,968 (15 %)	Asia: 64,073 (3 %)
	Sales	**EU: 59.4 %**	**EU: 84.1 %**	**EU: 64 %**
		North America: 8.5 %	Latin America: 6.6 %	North America: 9.5 %
		Asia: 18.6 %	Rest of the world: 9.1 %	Rest of the world: 26.2 %
2009	Production	**EU: 3,612,380 (60 %)**	**EU: 2,146,330 (71 %)**	**EU: 1,338,767 (54 %)**
		America: 1,151,568 (19 %)	America: 199,700 (7 %)	America: 831,284 (34 %)
		Asia: 1,243,572 (20 %)	Asia: 608,753 (20 %)	Asia: 110,808 (5 %)
	Sales	**EU: 54.1 %**	**EU: 80 %**	**EU: 60.7 %**
		North America: 7.1 %	Latin America: 8.5 %	North America: 10 %
		Asia: 25.4 %	Rest of the world: 11.5 %	Rest of the world: 29.3 %
2010	Production	**EU: 4,109,505 (56 %)**	**EU: 2,343,548 (65 %)**	**EU: 1,254,324 (52 %)**
		America: 1,341,349 (18 %)	America: 272,858 (8 %)	America: 865,823 (36 %)
		Asia: 1,692,517 (23 %)	Asia: 879,055 (24 %)	Asia: 133,208 (6 %)
	Sales	**EU: 49.6 %**	**EU: 85.2 %**	**EU: 60.5 %**
		North America: 7.6 %	Latin America: 6.7 %	North America: 3.1 %
		Asia: 30 %	Rest of the world: 8.2 %	Rest of the world: 36.6 %

Source OICA correspondence survey (2007, 2008, 2009, 2010), Volkswagen (2008, 2009, 2010, 2011), PSA Peugeot Citroen (2008, 2010), Fiat (2008, 2009, 2010), modified by the author

Since the decision-making structure of the EU is based on the codecision between the Commission, the Parliament, and the Council, it is important that none of these organizations can be entirely ignored. I argue that this dynamics in turn made 'institutional stickiness'[14] in the EU when it comes to the negotiations on fuel economy regulations. As a result, the institutional stickiness of the EU decision-making process further enabled the integration of a 95 g/km target by 2020, along with the 120 g/km target. This section describes the decision-making process of the European fuel economy regulations for 2015 and beyond. To do so, it looks at the respective decision-making process and main discussions that took place at the European Commission, the Council, and the Parliament.

The main actors in the European car climate policy include the European Commission, the European Parliament (Commissioners for the Environment and the Commissioner for Industry and Entrepreneurship), the European Council, national governments, ACEA, and the car manufacturers in each country. Figure 3.2 shows the decision-making process of the European fuel economy regulation implementation.

The process of the EU decision-makings on fuel economy regulation was characterised by dynamic, diversified actors with different degrees of authority given at different levels. As an executive body, the authority to propose the new fuel economy standards at the EU level is conferred to the European Commission (The European Commission n.d.). The Commission comprises of 27 members that are selected from one per each member state, who are bound to represent the interest of the EU as a whole rather than their home states. This implies that once the Commission adopts the target for fuel economy regulations, which was originally proposed by a member states driven by their own interests, the target itself persists for remainder of the EU decision-making process in order to deliver the interest of the EU as a whole; and this is where the institutional stickiness of the EU decision-making appears, the target of 120 g/km itself has not been changed since it was first proposed 13 years prior to the legislative process.

The proposal made by the Commission is simultaneously passed to the European Parliament and the European Council. The European Parliament and the European Council function as the legislative body of the European Union. The European parliament (the Parliament, hereafter) comprises of directly elected officials. The Parliament suffered from a lack of its presence in European decision-making process, and question of democratic legitimacy or a 'democratic deficit' was concerned (Crombez 2003; Moravsik 2008). Nonetheless, accompanying with progresses of European integration over years, there have been improvements in "the democratic legitimacy of the institutional system by reinforcing the powers of Parliament with regard to the appointment and control of the Commission and successively extending the scope of codecision procedure" (Europa n.d.). The codecision-

[14]'Institutional stickiness' refers to "the ability or inability of new institutional arrangements to take hold where they are transplanted" (Boetike et al. 2008, p.332). The concept is closely related to one of the conventional approaches in political science called 'historical institutionalism'. The approach focuses on institutions in order to find sequences of social, political, and economic behaviours of actors, i.e. how certain choices of actors in past cultivates present institutional arrangements (so-called 'path dependence') and how past institutional arrangements 'lock-in' to present institutions (so-called 'lock-in effect') (Steinmo et al. 1992).

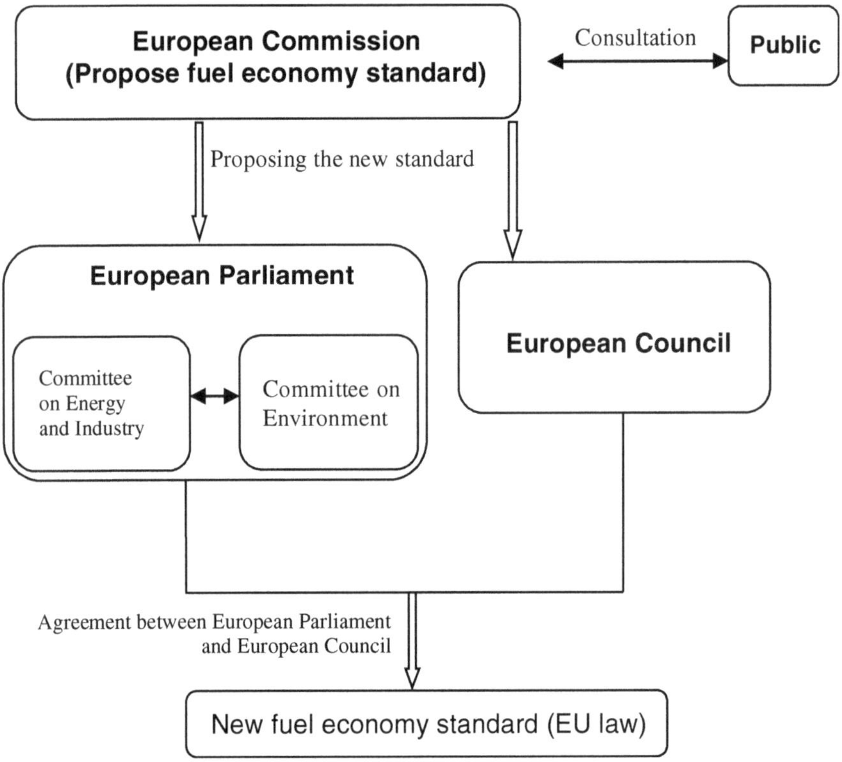

Fig. 3.2 European fuel economy regulation. *Source* Iguchi and Hillman (2012)

procedure was introduced by the Maastricht Treaty in 1992 and later the Treaty of Amsterdam (1997) and the Treaty of Niece (2003) expanded the number of legal bases where the procedures are applied. The procedure allows the Parliament, together with the Council of the European Union, has to approve the EU legislation that is proposed by the Commission. This, in turn, created a space for the environmental pressure groups to push for higher environmental standards, including stringent fuel economy regulation (the sections to follow describe how environmental NGOs influenced the discussions). In the European Parliament, committees are key actors in the adoption of EU legislation (Archick 2013). Following two committees were particularly played a critical role in shaping the Commission proposal on new fuel economy regulation: the Environment, Public, Health and Food Security (the environment committee, for short) and the Committee on Industry Research and Energy (industry committee, for short). Both committees appoint 'rapporteur' to draft on the legislative proposal under consideration, in order to draw up, amend and adopt legislative proposals. The committee then discuss, vote on, and amend the draft prior to present at a plenary session of the entire Parliament. In a plenary session, the committee's report is put to a vote and the Parliament adopts its position by a simple majority.

After agreement has reached in the Parliament, the Council of the European Union (the Council, hereafter) later adopts its position by a qualified majority. If the Council approves the Parliament's position, then the Commission proposal is issued as EU law. The Council is composed of 27 national ministers, each selected per member states. The discussions in the Council on the fuel economy regulations were characterised by the clash of interests between member states that on the one hand supports for a stringent fuel economy targets and those who attempted to weaken the targets of 120 and 95 g/km.

What is apparent in the decision-making process was the split between Germany and France. Much of the discussions in the Commission, the Parliament, the Council, as well as at the industry level, were marked by the divide between France and Germany. Virtually every country in the EU saw the automotive industry as a strategic important tool. With this reason, the apparent 'split' among the European car industries were caused and led the heavy lobbying endeavours exerted by the German car industry, together with its home state. Two factors can be identified as causes of this split. The first was that, as mentioned, the French and Italian car industries produced relatively light, small, and compact fuel-efficient vehicles, while the German (and perhaps British) car industry produced heavy, large, luxury vehicles with low fuel efficiency; because of this difference, the burden of sharing the emissions reduction target was a central discussion.[15] Second, the EU introduced a flexible credit-trading system called the 'Carbon Allowance Crediting System' for the achievement of the 120 g/km target. It allowed the manufactures to pool their emission standards with other manufacturers and gave credits to manufacturers for exceeding their target, allowing them to sell these credits to others who were below the target. This suggests that those who were producing light and highly efficient vehicles would benefit from this system, while the others would not. Consequently, it suggested that the German car manufacturers, with an average of 165 g/km in 2008, must buy credits from French and Italian manufacturers to achieve the legally-binding target.

In addition, the split affected other countries, especially those with manufacturing plants of the warring companies. Because of this dependency, there was a real reluctance at the national level for countries to displease the major players in the motor industry. A good example of this was Poland, which received no benefit whatsoever from opposing the emission regulations, given their low consumption rate of new cars. Still, Poland strongly opposed the new regulations. This could be because they wanted to strengthen its economy through new manufacturing plants.[16]

The following sections will look at detailed decision-making processes in each supranational institution.

[15]The French and Italian automobile manufactures were about to clear the 140 g/km voluntary target in 2008. In contrary, the average fuel economy of German automobile manufactures was 180 g/km in 2006, and 165 g/km in 2008. See T&E (2009).

[16]Based on interview with Mr. Greg Archer, T&E.

Commission Consultation Process

The European Commission was granted the ability to develop environmental initiatives beginning with the signature of the Maastricht Treaty in 1992 (Leveque 1996). This applied to the fuel economy regulation for passenger cars. Right after the Commission announced the mandatory reduction of CO_2 emissions December 1995, it held a public consultation on the implementation of the renewed strategy to reduce CO_2 emissions from passenger cars and light commercial vehicles from 15 March to 15 July 2007. In this process, 28 views were expressed by citizens, and 41 views were expressed by several organizations, including the ACEA, the JAMA, General Motors (GM), KAMA, the German association of the automotive industry (VDA), and several environmental NGOs including T&E and Greenpeace (European Commission 2007d).

While the environmental NGOs, such as T&E, WWF, FoE, and Greenpeace, asked for an improvement in the fuel economy at a rate of 5 % a year, achieving 120 g/km by 2012, 80 g/km by 2020, and 60 g/km by 2025 (Friends of the Earth 2007; Greenpeace 2007; T&E 2007b; WWF 2007), the industry attempted to weaken these standards and timetables. Notably, one the one hand, the ACEA generally supported further reduction in CO_2 emission from passenger cars and accepted the EU's target of 120 g/km target itself, but on the other hand, it opposed the implementation of the regulation by 2012, since "the proposed 2012 date is unrealistic given industrial process in the automotive industry, and the lack of planning certainty in the absence of a legislative framework…limited potential for further changes exist" (ACEA 2007a). Instead, it insisted that 2015 was the earliest possible date for implementing the car technology target of the CO_2 legislation. The president of the ACEA, Sergio Marchionne, said, "the feasible date for implementation of new legal requirements is 2015" (Automotive Engineer 2007). Besides, it argued that 135 g/km should be the target for car/engine technology and that the rest of 15 g/km should be achieved through complementary measures, such as bio-fuels, eco-driving, and infrastructure measures.

The voice of ACEA is considered to reflect the preferences of the VDA. It argued that 2015 is the key date for the realistic, step-by-step implementation of a future CO_2 regulation (VDA 2007). Most importantly, it argues for a differentiated, weight-based target, and argues against a unified CO_2 objective for the new car fleets of all manufactures given that "it would not take account of differing customer requirements, usage requirements or technic al interdependencies" (VDA 2007). Foreign manufacturers, such as JAMA and KAMA, harmonized their positions on new regulatory fuel economy standards with the ACEA. JAMA emphasized that 'the proposed new target application year and value should be reconsidered'. Namely, it supported the year 2015, which was identical to the Japanese fuel economy standards. For the target value, it supported, in line with ACEA, 135 g/km to be achieved by vehicle technologies and 15 g/km by complementary measures. Moreover, it supported the weight-base system as introduced in Japan (JAMA 2007). Similarly, KAMA supported 135 g/km through vehicle technology improvements and 15 g/km to be fulfilled by other measures. Besides, it emphasized that the target year should be extended to 2015 (KAMA 2007).

The US manufactures, such as GM, also supported 'an ambitious' technology target of 135 g/km for 2015, in agreement with ACAEA, JAMA, and KAMA (General Motors 2007). This means that all automobile industry, regardless to European, Japanese or American, harmonized their positions to support weaker regulatory target, that is, a target of 135 g/km by 2015. For the European industry, the industry network was relatively strong for influencing EU policy-making. When it came to lobbying the Commission, the Parliament, and the Council, the manufacturers exerted their influence through the ACEA network because the ACEA was more effective than the lobbying of individual companies.[17] For the very same reason, it is assumed, that foreign automobile industry harmonized their position with the European industry in order to enhance its leverage against the stringent European regulations.

Despite automobile industry's opposition to more to stringent fuel economy regulations, the European Association of Automotive Suppliers (CLEPA) supported a stringent target of 120 g/km by 2012. CLEPA supported the stringent fuel economy target as its membership perceived the research and innovation opportunities and the environmental benefits of deploying low-carbon technologies on the market.[18] Hence, this situation highlights the diverging positions of the vehicle manufactures and the automotive suppliers.

To sum up, the industry preferences of an integrated approach were reflected in the Commission proposal, which proposed the EU objective of achieving 120 g/km focusing on a mandatory reduction of the 130 g/km for the average new car fleet by improvements in vehicle motor technology and a further reduction of 10 g/km by an 'integrated approach' by 2012 (European Commission 2007b).[19] Still, what was also important is that the Commission's proposal did mention the "possibility of setting more ambitious objectives beyond the current Community target of 120 g/km at a later stage", and supported research efforts toward reaching a 95 g/km target (European Commission 2007a, b, c, d). The next section examines how the Commission proposal was discussed in the European Parliament.

---

[17]Based on interview with Mr. Petr Dolejsi, Mobility & Sustainable Transport Director, European Automobile Manufacturer's Association (ACEA). Interview conducted at Brussels, Belgium (16 May 2012).

[18]Based on interview with Mr. Pierre Laurent, Senior manager, Technical Department, European Association of Automotive Suppliers (CLEPA). Interviewed at Brussels, Belgium (22nd May 2012).

[19]The 'integrated approach' was one of the key strategies for the European car industry (ACEA 2007b). Since the 'integrated approach' reduce burdens on the car industry to purely pursing on technological innovations by giving spaces to the introduction of bio-fuels, eco-driving and model shifts, this approach serves the interest of the European car industry. This approach has even appeared in the final report released by the Competitive Automotive Regulatory System for the 21st Century (CARS21), which was launched by the European Commission aiming to "make recommendations for the short-, medium-, and long-term public policy and regulatory framework of the European automotive industry" (CARS21 2006). Actually, 'integrated approach' was also promoted by the Japanese automobile industry at the APP (see Iguchi (2012) for details). This suggests that integrated approach was employed between Japanese, European, and American automobile industry to avoid the cost of raising fuel economy regulation elsewhere.

Discussion in the European Parliament

The European Parliament had been showing its support for the stringent fuel economy regulations from the very beginning of the regulatory process. In 1997, prior to the ACEA-Commission voluntary agreement, the Parliament strongly pushed for the very stringent target of 90 g/km by 2007 and urged the Commission to abandon the voluntary agreement by the industry in support of mandatory regulations (The ENDS Report 1997). In 2005, it called for "a policy of strong measures to reduce emissions from transport, including mandatory limits for CO_2 emissions from new vehicles in the order to 80–100 g/km for new vehicles in the medium term to be achieved through emissions trading between car manufactures" (European Parliament 2005). Furthermore, in the report adopted by the European Parliament on October 2007, it showed its support for the Commission's proposal, mentioning a 95 g/km target by 2020 and a 70 g/km target by 2025 (European Parliament 2007a). But, despite that fact that such a position had been adopted by the Parliament, the discussions in the Parliament during 2007 and 2008 were characterized by the division between the Environment, Public Health, and Food Security Committee (the environment committee, for short) and the Committee on Industry Research and Energy (industry committee, for short).

On the one hand, in an opinion report released on 17th June 2007 by Rebecca Harms, a rapporteur of the industry committee, offered support for additional complementary measures to be taken to fulfil the 120 g/km target. She called on the Commission to "take into account of technical feasibility, cost-efficiency, environmental impact and affordability over the vehicle life-cycle of new cars when adopting any binding legislative measures which might influence the pace of vehicle fleet renewal" (Committee on Industry Research and Energy to the European Parliament 2007). On the other hand, an opinion report published on 24th September 2007 by Chris Davis, a rapporteur of the environmental committee, criticised the Commission's proposal that set 10 out of 120 g/km by 2012 to be achieved by complementary measures. He claimed that this complementary measure "has reduced the clarity of the target and encouraged some manufactures to think that they can use bio-fuels as a means of avoiding significant design changes" (European Parliament 2007b). Furthermore, he argued for much more stringent targets than the proposal made by the Commission, stating that the timetable for 120 g/km target should be by 2012, as well as insisting that average emissions from passenger cars should not exceed 95 g/km by 2020 and that a further reduction to 70 g/km be achieved by 2025 (Committee on the Environment Public Health and Food Safety to the European Parliament 2007).

Apparently, these contradicting proposals within the Parliament clearly reflected the views of automobile industry and the environmental NGOs. As previously mentioned, the industry argued that 135 g/km should be the target for car/engine technology by 2015 and that the other 15 g/km should be achieved through complementary measures, such as bio-fuels, eco-driving, and infrastructure measures. The environmental NGOs, however, supported improving fuel economy at a rate of 5 % a year, achieving 120 g/km by 2012, 80 g/km by 2020, and 60 g/km by 2025.

In the subsequent discussions that took place on 1st September 2008, the industry committee opposed the Commission's proposal of a 130 g/km target with a 10 g/km complementary measure by 2012. This meant that industry clearly did not want what the Commission proposed. Instead, in an opinion report released on 3rd September 2008 by Werner Langen, a rapporteur of the industry committee, he emphasized the following three points from the perspective of the competitiveness within the car industry. Firstly, he argued for a postponement of the target year from 2012 to 2015, where the industry should meet a gradual fulfilment of the standard year by year; e.g., 60 % of its fleet by 2012, 70 % by 2013, 80 % by 2014, and 100 % by 2015. Secondly, although the Langen report mentioned that the 95 g/km target should be the aim for 2020, it emphasized that this long-term target should be established by taking an impact assessment into account. This left room for change, allowing the industry can exert its influence according to the content of this proposed impact assessment. Third, it asked for flexibility, such as the introduction of a 'pooling system' (manufacturers could form a pool for the purpose of meeting their obligations) and promotion of alternative-fuel, low-emission vehicles and zero-emission vehicles on the EU market by counting new passenger cars with less than 50 g/km and each alternative-fuel vehicle as 1.5 cars and every zero-emission vehicle as three cars up to and including 2015 (Committee on Industry Research and Energy to the European Parliament 2008).

In reaction to the Langen report, the environment committee drafted its own report. A report drafted by Guido Sacconi (environment committee, published on 8th May 2008) reemphasized the need to achieve a 120 g/km target by 1 January 2012, as well as to set a target for the new car fleet of average emissions of no more than 95 g/km from 1 January 2020; it also asked the Commission to present a proposal for regulations setting the 95 g/km target for 2020 by 31st December 2014 (Committee on the Environment Public Health and Food Safety to the European Parliament 2008).

The final agreement in the Parliament was delivered on 17th December 2008, when the Parliament adopted its position paper by 449 votes against 98 votes (European Parliament 2008). In the final agreed text, proposals made by the industry committee were reflected to a large extent: The Parliament adopted the 120 g/km target by the year of 2015, 130 g/km to be achieved by means of improvements in vehicle motor technology and 10 g/km to be achieved by the integrated approach. Regarding the target year, it supported the 'phase-in' timetable in which the industry had to fulfil the 130 g/km target by 65 % in 2012, 75 % in 2013, 80 % in 2014, and 100 % from 2015 onwards. Furthermore, several flexibility measures such as 'super credits,'[20] a 'pooling system,'[21] and

[20]It aims to encourage manufactures to produce low-carbon vehicles, by counting new cars of less than 50 g/km as 3.5 in 2012, 3.5 in 2013, 2.5 in 2014, 1.5 in 2015, and 1 car from 2016 on. This means one sales of new car less than 50 g/km counts as 3.5 cars sold in 2012.

[21]Under this system, manufacturers can form a pool for the purpose of meeting their obligations.

'eco-innovation'[22] were also mentioned in the agreed text. Although much of argument made by environment committee was not reflected in the final text, it was notable that the Parliament noted the 95 g/km target for 2020 and urged the Commission to complete a review of this by 1st January 2013. What made the Parliament to note 95 g/km target for 2020? According to interview conducted to European environmental NGOs, which had lobbied the Parliament to push very hard for a stringent 95 g/km target, this is because all of the industry's attention went to weakening the target of 120 g/km by 2015 by including integrated approach.[23] Consequently, the target of 95 g/km by 2020 was raised at the negotiation table and simply passed without many comments. As a result, what happened was that industry has to work to surpass the 2015 target, which they worked so hard to undermine, in order to achieve the 95 g/km target. The next section looks at discussions in the Council of Ministers, where the preferences over both 120 g/km and 95 g/km targets were contested between Ministers of each member state.

Discussion in the Council of Ministers

The discussions of the Council of the EU, which represents the executives of the EU member states, on the Commissions' proposal started in the 2785th Council Meeting held on 20th February 2007 (Council of the European Union 2007b). The overall Council conclusion in the report released on 3rd July 2007 reconfirmed its support for the target of 120 g/km by 2012, 130 g/km to be achieved by technological improvements and 10 g/km to be achieved by additional measures. Hence, it supported the proposal made by the Commission, recognizing that the 'European car industry can gain significant first mover advantage through research and development promoted by ambitious CO_2 reduction targets and new environmental technologies' (Council of the European Union 2007a). Furthermore, at its policy debate on the 2856th Council meeting held on 3rd March 2008, several conclusions were made. First, it reaffirmed that the emissions reductions from the road transport sector should contribute to the Community's overall objective of limiting the global temperature increase to 2 °C. Second, delegations supported the integrated approach as complementary measures to achieve the 120 g/km target, which was to be fulfilled by 130 g/km by technological improvements and 10 g/km by the integrated approach. Third, it emphasized the need to balance the competitive issues and the need to reduce CO_2 emissions. Finally, several delegations mentioned the need for a long-term target as a signal to the industry (Council of the European Union 2008a).

The divergence of opinions over the fuel economy regulations appeared in the presidency progress report released on 20th May 2008. It reported that some

[22]It allows manufacturers to use innovative technologies from parts suppliers to achieve their specific emission targets of up to 7 g/km.

[23]Based on interview with Mr. Greg Archer, T&E.

divergences were identified in the following five areas: the utility parameters of vehicles (whether the regulation should be based on weight of vehicles), slope of the curve (the percentage of the CO_2 emission reduction target), penalties for incompliance, calendars (target year), and the long-term target. First, regarding the discussion on utility parameters, there was a division of preference between weight-based approach as the Commission had proposed and the footprint approach (vehicle size approach) that reasoned that a size-based approach would provide stronger incentives for manufactures to reduce the weight of cars. Second, the discussion on the slope of the curve was based on the Commission's proposal of a slope of 60 % (130 g/km). While some delegations supported a higher percentage of slope of curves (65–80 %, therefore more stringent than the 120 g/km target), other delegations preferred a slope lower than 60 % (20–30 %, therefore much weaker than the 120 g/km target). Third, while some delegations supported the Commission's proposal of gradual penalties in order to enhance industry's compliance to the regulation, others preferred to lower such sanctions. Fourth, while most of the delegations agreed to postpone the timetable for 120 g/km target by 2015, some preferred to start in 2015, while others supported the phase-in for the car fleet from 2012 to 2015. Finally, while delegations agreed to set a long-term target for 2020, the delegations were divided regarding whether to set it at 95 g/km or to avoid making reference to a concrete figure (Council of the European Union 2008c).

This divergence continued at the 2874th Council meeting, held on 5th June 2008. The debate in the meeting was based on the presidency progress report. The council focused on three aspects. The first aspect was the discussion of a utility parameter. Most of the delegations supported the weight-based approach, while some other delegations supported the footprint approach. The second aspect was the slope of the curve, namely, the percentage of the CO_2 emission reduction target for new passenger cars, where diverging views on percentage were still expressed. The final aspect was the penalties for manufacturers who failed to meet the 130 g/km target. Some delegations expressed lower sanctions or one with a certain degree of flexibility (Council of the European Union 2008b).

What is apparent from the member states' diverging opinions is that they were clearly divided between those who preferred stringent fuel economy regulations and those who wanted a rather weak, easily attainable target. In particular, some member states wanted more a stringent fuel economy target than 120 g/km, more penalties of incompliance, and a 95 g/km target set for 2020. It is notable that some member states also wanted a weak 120 g/km target with lower penalties of incompliance and no target for 2020.

Final Outcome

In December 2008, the 120 g/km target (130 g/km purely by technological developments) was agreed upon between the European Parliament and the European Council, and it was issued as EU law (Official journal of the European Union 2009).

Table 3.4 The milestones of European fuel economy regulations for 2015 target

Year	Policies relating to the fuel economy regulation	Target year
1991	ACEA announced its voluntary target of 10 % reduction	1993–2005
1995	The European commission announced the target of 120 g/km/km	2005
1999	A voluntary target between ACEA and the European committee with the target of 140 g/km	2008
2007	The European commission announced to introduce new legally binding CO_2 target for passenger cars	2012
2008	A conditional 120 g/km target was agreed between national governments and the European Parliament	2015

Source created by author

The Commission originally proposed this restriction in 1995, so it took 13 years to finally deliver a mandatory fuel economy standard, as can be seen in Table 3.4.

In its final outcome, the 120 g/km target was agreed upon, although on a conditional basis. The conditions included the use of the proportional increase of average fuel economy, of 65 % in 2012, 75 % in 2013, 80 % in 2014, and 100 % from 2015 onwards. Therefore, the views expressed by the Industry Committee of the Parliament were reflected in the final agreed text. Moreover, the industry preference of the 'integrated approach' was incorporated into this target. The European car manufacturers were required to achieve 130 g/km through vehicle technology improvements and to achieve the remaining 10 g/km through complementary measurers such as the increased use of bio-fuels, traffic road-safety management, fuel-efficient tyres and air conditioning, and changes in driver behaviour.

Hence, much of the industry's preferences were reflected in the final outcome. In addition, the other industry preferences, such as eco-innovations and super credits, appeared in the final text. The industry wanted these options included to reduce the stringency of the requirements. Many of these ideas came from Germany for two reasons. First, it is important to note that the French government was currently filling the role of the EU Presidency. Therefore, France was managing the negotiations rather than actually participating in them. Therefore it is reasonable to assume that France was forced to accept a lot of the German proposals. Second, was a very high level and bilateral political deal done by Sarkozy and Markel between Germany and France.[24] Essentially, what was never made public was that the countries agreed that Germany would mainly get what it wanted regarding the automobile industry and CO_2 regulation and that France would mainly get what it wanted in a future airbus deal.

Within the industry, the German industry has the strongest voice within the ACEA, followed by the French and Italian industries.[25] Moreover, the German voice was reflected overall in the strategies taken by the ACEA. According to an

[24]Based on interview with Mr. Greg Archer, T&E.

[25]Ibid.

interview conducted with the ACEA, "there were diverging positions between European car manufactures".[26] Yet, ACEA had chosen the weight-based approach, although there was discussion about a unified target because under the weight-based approach. Further, the German car industry emphasized the importance of eco-innovation credits, which was reflected in the final outcome. In addition to the eco-innovation credit, the phase-in and delay to fulfil the target were also strongly emphasized by Germany, while France did not appear to make any contribution to the agreement. The German manufactures insisted on the introduction of eco-innovation credits because they did not want to have to make big changes on their engine and vehicle weights; rather, they wanted to be able to make changes a little at a time, such as on the lights, air-conditioning, tires, and consumer and driver information monitoring. Simply improving engines and vehicle weight, for instance, costs much more than improving the lights or air-conditioning. Furthermore, German manufacturers pushed for weight-based standards, while the French industry pushed for the size-based or footprint approach, especially Renault.[27]

While many of the industry preferences were reflected in the final outcome, the target of 95 g/km was set, which came about because of the global competitiveness of the auto industry. The EU is the first among the three studied cases to propose the stringent target for 2020. First of all, according to the interview conducted with a Policy Officer at the European Commission, although there was no policy coordination between Japan and the United States, the fuel economy regulations in those countries were taken into consideration when developing the proposal:

> I don't think there was a thing that can be called as 'policy coordination' between the EU and Japan… In terms of coordination, I think the important point is of course we are aware of what is being done in Japan and the US; and I am sure Japan is aware of what the EU is doing. So I think the competition is perhaps not the right term, but there is certainly an awareness of what is happening in the other parts of the world helps to make what we are doing more palatable for our politician. I am sure that it is same for people in the US and Japan, that when you can see that other regulatory jurisdictions are taking similar actions at similar rates, its strange to do it yourself.[28]

Unlike the 120 g/km target, when first proposed, the 95 g/km target is based on calculations. According to an interview conducted with the Netherlands Organization for Applied Scientific Research (TNO), which is one of the key research institutions involved in assessing possible targets and other modalities for the passenger car CO_2 regulations for 2015 and 2020, an preliminary assessment of possible target levels for 2020, ranging from 85 g/km to 105 g/km, was carried out in 2009 using indicative cost curves for 2020. A more detailed assessment of the impacts of a 95 g/km, using new cost curves for 2020, was carried out in 2011.[29]

[26]Quoted from interview with Mr. Petr Dolejsi, ACEA.

[27]Based on interview with Ms. Franziska Achterberg, Greenpeace EU.

[28]Quoted from interview with a Policy Officer, European Commission.

[29]Based on interview with Dr. Richard Smokers, TNO Delft. Interviewed at Delft, Netherlands (24th May 2012).

Assumably, the target of 95 g/km was picked as a compromise between various actors with various interests at various levels, ranging from the various directives in the Commission, the environment committee and the industry committee in the European Parliament, and the diverging preferences of the Council of Ministers. In particular, European environmental NGOs pointed out that France has pushed for 95 g/km target for 2020.[30] The details of why France has pushed the 95 g/km forward is yet to be known; however, given the high achievement of fuel efficiency in French auto industry, we can assume that it supported 95 g/km target in order to enhance competitiveness against German auto industry, whose preferences were reflected in the discussions on the 120 g/km target to a great extent.

Another question is, in deciding 95 g/km, do fuel economy regulations in Japan and the US – which of these are also the main car manufacturing nations – influence the standards? According to TNO:

> So far they have not, but for the 95 g/km target and beyond they (the competitiveness issues) will influence the legislation. In the Impact Assessment report, that we helped the Commission to write, one chapter is competitiveness impacts. Especially there, it is very important to know what other regions are doing. The Commission has asked us to look at the Japanese and US fuel economy regulations at least to see if we could copy some ideas from that, but the 2020 target is set without consideration to the other countries. For post 2020 targets the issue of competitiveness will become more important, though. If the target in Europe is much stricter than in other regions, the European manufactures have to do much more in home markets than in other markets so it will influence their competitiveness. Whether these impacts would be positive or negative is to yet be determined.[31]

Furthermore, environmental NGOs have observed that the converging trend is primarily a competitiveness issues. They argue that if Europe develops technologies to make its vehicles more fuel efficient, then those technologies are equally applicable in the US market or Japanese market or Chinese market.[32] An expert on European fuel economy regulation at Greenpeace point out, "in the light of what Europe has already agreed, the US followed to enhance its fuel economy regulation, not so much for the environmental reason, but for the sake of the enhancing competitiveness of the industry. Therefore even though there was not so much coordination, the EU's target for 2015 and 2020 triggered other regions to enhance their fuel economy regulations".[33]

Overall, regarding the policy-making of the European fuel economy regulations, competitiveness and climate change were the prime reasons for setting the 95 g/km target and beyond. It follows that the European fuel economy regulations has been formulated as a result of a clash of interests among divergent actors at different levels. While the industry strongly resisted any increase of fuel economy standards, environmental NGOs acted as 'norm entrepreneurs' to push the 120 and 95 g/km

[30]Based on interview with Ms. Franziska Achterberg, Greenpeace; Interview with Mr. Greg Archer, T&E.

[31]Interview with Dr. Richard Smokers, TNO Delft.

[32]Based on interview with Mr. Greg Archer, T&E.

[33]Based on interview with Ms. Franziska Achterberg, Greenpeace EU.

targets forward. These targets were then presented on the negotiation table at the EU level. Once these targets were put before the EU, they became very hard to remove; hence, the 'stickiness' of EU institutions helped the legalizations of these targets.

Target for Beyond 2015: Legalizing 95 g/km Target

The European Commissions' draft proposal on means to achieve the 95 g/km target was issued on 11th July 2012. Although weight-based approach continued as the basis of the regulations, it limited the use of super-credits, by allowing cars emitting below 35 g/km could only get super-credits between 2020 and 2023 with a multiplier of 1.3 for a maximum of 20,000 vehicles per manufactures (European Commission 2012).[34]

An Ambitious target was further proposed by the Committee on the Environment, Public Health and food Safety in April 2013. It not only confirmed the Commissions' 95 g/km target by 2020, but also proposed to set a range of 68–78 g/km target for 2025 (Committee on the Environment Public Health and Food Safety to the European Parliament 2012).

However, as it confronted strong oppositions from Germany,[35] the Commissions' proposal was weakened in the final agreement document released on 11th March 2014 (Official journal of the European Union 2014). Although it confirmed 95 g/km target by 2020, the indicative target for 2025 disappeared from the text, and only called for the Commission to propose beyond 2020 target by 2015. Furthermore, it set that 95 g/km target to be met by the end of 2020 onwards, after a phasing-in period in which 95 % of new cars must respect the target by 2020. In addition, the use of super-credit is expanded compared to the Commission' proposal.

3.7 Summary

This chapter has looked at how Europe's climate policies for car CO_2 emissions have been constructed. Milestones and key factors of the EU fuel economy regulation are summarized in Table 3.5. The European car fuel economy regulations emerged out of the aim to limit growing CO_2 emissions from the road transport sector in the early 1990s. Originally, the European Council requested the Commission to legalize its first fuel economy regulation by 1992, the year of the

[34]This means manufactures can count the sales of one vehicle emitting below 35 g/km as 1.3 vehicles.

[35]For instance, it proposed a phase-in on the stringency of the regulations, where 80 % of new cars must meet 95 g/km by 2020, and fully meet by 2024 (T&E 2013). In other words, the proposal delays the target being met four years later, in 2024, not 2020.

Table 3.5 The milestones and key factors of EU fuel economy regulation

Year/ Event	Milestones of EU fuel economy regulations	Key factors
1990s	**1991**: the commission started to formulate fuel economy regulations	Reduction of CO_2 as the principal motivation for the commission
	1991: ACEA's self-regulation to reduce 10 % between 1992 and 2005	120 g/km target emerged out of competitiveness concern of German auto industry
	1994: Germany proposed '120 g/km target by 2005'	Conflict between Germany and France caused ACEA to act as laggard
	1995: the commission proposed '120 g/km target by 2005'	A voluntary agreement was formed between ACEA and the commission as a compromise
	1995: ACEA lobbied to postpone the target year by 2010	
	1997: ACEA offered to improve '155 g/km by 2005'	
	1998: A Voluntary Agreement of '140 g/km target by 2008'	
2000s	**2007**: the commission proposed mandate reduction of '120 g/km (130 g/km) by 2012' and possibility to research towards '95 g/km by 2020'	NGOs revealed that industry did not make progress towards the voluntary agreement
	2008: the parliament adopted '120 (130)/km by 2015', and urged the Commission to review '95 g/km by 2020' target by 2013.	NGOs pushed 95 g/km target through environment committee in the parliament
	2009: '120(130)g/km by 2015' and '95 g/km by 2020' targets set by the EU law.	Industry successfully lobbied the commission, the parliament, and the council to reflect its preference
		95 g/km target is to enhance competitiveness of auto industry; France in particular pushed for 95 g/km target
2010s	**2012**: the commission put forward a proposal how the 2020 target to be met	While the 95 g/km target agreed, many flexibilities measures were allowed as a result of German lobbing
	2013 parliament confirmed 95 g/km target and proposed 68–78 g/km target by 2025	
	2014: 95 g/km target set by the EU law, while allowing many flexibility measures	

Source created by author

adopting the UNFCCC. Although the Commission could not meet this request due to divergence of preferences among member states and their industry, concerns for climate change issue kept continue to push the European fuel economy regulations. Just before the adoption of the Kyoto Protocol in 1995, Germany proposed the

'120 g/km by 2005' target. This target was mandated prior to COP15 in Copenhagen in 2009, in which parties to the UNFCCC were expected to adopt new protocol to limit global CO_2 emissions for the mid-term target. Moreover, the 95 g/km target for 2020 was agreed among member states simultaneously with the 120 g/km target in 2009. Currently, the 95 g/km target is under the process of legalization, in consistent with the EU's international climate policy for 2020.

Despite concerns on industry competitiveness issues, business in general lobbied to delay the introduction of the 120 g/km target from its very beginning up until now. Throughout the EU fuel economy regulation process, the industry exerted its influence to a large extent on each EU institution. Such reluctance of the industry is rooted in conflicts between German and French automobile industry. On the one hand, Germany insisted the introduction of a weight-based system, suggested that fuel economy regulation could be differentiated according to the weight of vehicles, as long as the mean of all vehicles sold in Europe satisfied the 120 g/km target. This meant that the French and Italian manufacturers, which produced lighter cars, would have to carry a heavier burden for improving fuel efficiency than the German manufacturers. On the contrary, French and Italian automobile industry wanted an absolute target of the 120 g/km for every manufacturer, regardless of the weight of the vehicle. They preferred a regulatory incentive that would give financial benefits to achievers and fines to underachievers. This, in turn, meant that the German manufacturers would have to take on considerable burdens to satisfy the target. The conflict resulted in business lobbying activities through ACEA, a network of European automobile industry over the EU decision-making process. The position of ACEA seem to reflect the lowest common denominator between the two, thus it is natural to assume that lobbying to weaken and delay the supranational target would benefit both German and French automobile industry, since German automobile industry would not want a stringent target, while French industry would not want a stringent target based on weight-based regulations. This is particularly evident in the discussions in the European Parliament during 2007–2008, the industry committee, which reflect preference of automobile industry, opposed to the Commission proposal of the 120 g/km target, arguing for an introduction of 'phase-in timetable' to delay the target year and 'flexibility mechanisms' to lighten the burden of automobile industry.

The critical junctures were brought by environmental NGOs. In 2007, environmental NGOs started to publicize the performance of each manufacturer to show that they were not respecting or striving toward the proposed target. As a result, it revealed that the industry efforts to improve fuel economy were way below the standards set by voluntary agreement, and thus directly led the Commission to legalize the 120 g/km target. This means that the environmental NGOs were instrumental to promote normative change that directed the Commission to propose the legislation, and hence functioned as a 'norm entrepreneur'. Furthermore, environmental NGOs were key to push for the stringent '95 g/km by 2020' target in the Parliament. In the Parliament discussions, the environment committee strongly claimed that timetable for the 120 g/km target should be by 2012, as well as insisting that average emissions from passenger cars should not exceed 95 g/km by

2020 and that a further reduction of 70 g/km should be achieved by 2025. Consequently, although proposal of industry committee on 120 g/km did win over the environment committee in the final agreement, the Parliament urged the Commission to review the '95 g/km by 2020 target' by 2013. According to interview conducted to European NGOs, the appearance of the 95 g/km target was possible with a lot of encouragements from environmental NGOs. This suggests that if the Parliament strengthens its role in the EU decision-making process over time, there would be more channels available for the environmental NGOs to exert influence over the EU decision-makings.

What also made the 95 g/km target possible was the institutional stickiness of the EU decision-making process on fuel economy regulation, which can be characterised as a supranational codecision-making process between the Commission, the Council of the EU and the European Parliament. Because an authority to propose new regulatory standards is conferred to the Commission, which is an executive body that represents the interest of the EU, not each member states, the stringent 120 g/km and the 95 g/km target were proposed in order to reduce CO_2 emissions from road transport sector as well as to enhance competitiveness of European automobile industry. The proposal made by the Commission then passed to both the Council and the Parliament in parallel. Although national interest, along with business and NGOs lobbying excreted in the Parliament and the Council, and to an extent business lobbying undermined the 120 g/km target, however, the target itself remained for more than 13 years.

To sum up, growing environmental interests, especially by the European Commission beginning in the late 1980s, set a milestone for the EU's leadership. In particular, I have emphasised that civil society played a critical role as a 'norm entrepreneur' in this process. The fundamental rationale of the European Commission in proposing the stringent fuel economy regulations was to reduce the growing CO_2 emissions from cars as well as to enhance international competitiveness of its car industry. The civil society was critical in delivering the Commission's proposal. Coupled with Europe's complex supranational decision-making structure, the voices of civil society were well reflected as the European Parliament gained influence over the EU decision-making on fuel economy regulations. In contrast to the role played by civil society, the European car industry acted as laggard. It was characterised by the conflicts between the German and French car manufacturers. It has been argued that this conflict delayed the introduction of the Europe's fuel economy regulation for more than 13 years. European climate policies in the context of addressing CO_2 emissions from cars are constantly shaped and reshaped by the clash of divergent interests among its member states, mainly between France and Germany. It is especially notable that the EU's 120 g/km target was originally proposed by Germany in 1995, which, 13 years later, strongly opposed the legalization of the target. Table 3.6 summarizes these characteristics of EU fuel economy regulations.

The next question relates to the implication of the EU fuel economy regulations on the regulatory convergence. It is argued that the growing environmental and industry's competitiveness interests in Europe beginning in the mid-1990s, coupled with dynamic EU policy-making, led the EU to commit the highest fuel economy

Table 3.6 The characteristics of EU fuel economy regulation

Key factors	Description
Motive	To limit growing CO_2 emissions from the road transport sector and to enhance competitiveness of auto industry
Competitiveness issues	Targets are formulated in order to enhance industry competitiveness; Europe as fairly big market and European standard influence Japanese and US manufactures
Decision-making process	Co-decision procedure between the commission, the council and the parliament
Business	Split between Germany and France. Much of German preferences were reflected in the final outcome
NGOs	Played a critical role by revealing business achievement of the Voluntary Agreement; and pushed 95 g/km target forward
Critical Juncture	In 2007, pressures from domestic and abroad led the EU to legalize stringent fuel economy regulation

Source created by author

Table 3.7 Share of passenger cars in EU, by manufactures 2000–2010

Year/Country	Volkswagen (%)	Peugeot (%)	Fiat (%)	Ford (%)	Toyota (%)
2001	10	8	8	8	3
2002	10	9	6	9	4
2003	10	8	6	8	5
2004	10	8	6	8	5
2005	10	7	5	8	5
2006	11	7	6	8	6
2007	10	7	6	8	6
2008	11	7	7	8	5
2009	11	7	7	9	5
2010	11	7	6	8	4

Source created by author based on ICCT (2011)

regulations in the world. Its standards have an impact on any countries that export their cars to Europe or import European cars. Given the size of the European car market, which is the biggest among the three cases studies, is about two times larger than the US market and 13 times larger than Japanese market, the impact of the regulation on the foreign manufactures is also valid. Table 3.7 shows the share of passenger cars in the EU, both by European, Japanese, and American manufacturers. It is apparent that the share of both Ford and Toyota are as large as the European manufactures.

The influence of the European fuel economy regulations on these manufactures is also valid. Figure 3.3 shows a comparison of the CO_2 emissions from passenger cars by both European and Japanese manufacturers.

It is apparent that Toyota significantly improved its fuel economy during the voluntary agreement with the European Commission (2004–2006). Furthermore,

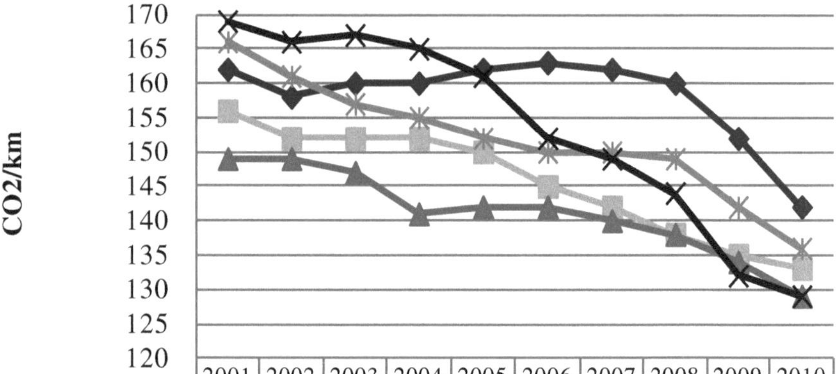

	2001	2002	2003	2004	2005	2006	2007	2008	2009	2010
Volkswagen	162	158	160	160	162	163	162	160	152	142
Peugeot	156	152	152	152	150	145	142	138	135	133
Fiat	149	149	147	141	142	142	140	138	134	129
Ford	166	161	157	155	152	150	150	149	142	136
Toyota	169	166	167	165	161	152	149	144	132	129

Fig. 3.3 Comparison of CO_2 emission from passenger car by manufactures, 2001–2010. *Source* created by author based on ICCT (2011)

when the EU's mandatory fuel economy regulations began to be discussed in 2007, all manufacturers started to improve their fuel economy. In particular, Toyota improved by 20 g/km between 2007 AND 2010, and Ford improved by 14 g/km in the same time range. This clearly demonstrates the impact of the European fuel economy regulations on Japanese and American manufacturers. This suggests that EU-level legislation automatically affects foreign car manufacturers who wish to sell their products in Europe. In other words, by having the highest fuel economy regulations in the world, Europe is likely to have the normative power, which, in turn, becomes the shared norms/international structure that influence the behaviours of the other countries. Thus, this spreads EU environmental norms and serves to legitimize EU rules prior to world trade agreements. The next chapter will look at the construction of Japanese fuel economy regulations, which, to a large extent, have been influenced by the European regulations.

References

ACEA (2007a) ACEA answer to commission consultation on CO_2/cars communication final. Brussels, 13 July 2007

ACEA (2007b) Press release—reducing CO_2 emissions: working together to achieve better results. Brussels

ACEA (2009) The automobile industry in Europe. Retrieved 10 Feb 2013, from http://www.acea.be/collection/the_automobile_industry_in_europe/

Automotive Engineer (2007) 2012 emissions date unfeasible, says ACEA. Automotive Engineer 32(6):14

Archick K (2013) The European parliament. Washington DC

Boetike PJ et al (2008) Institutional stickiness and the new development economics. Am J Econ Sociol 67(2):331–358

CARS21 (2006) A competitive automobile regulatory system for the 21st century. Reproduction, Brussels

Comission Staff of the European Communities (1998) Commission staff working paper: environmental agreement with the European automobile manufactures association (ACEA) on the production of CO_2 emissions from passenger cars. Brussels

Commission of the European Communities (1995) COM(95) 689 final Communication from the commission to the council and the European parliament: a community strategy to reduce CO_2 emissions from passenger cars and improve fuel economy. Brussels

Committee on Industry Research and Energyto the European Parliament (2007) Opinion of the committee on industry, research and energy for the committee on the environment, public health and food safety on community strategy to reduce CO_2 emissions from passenger cars and light commercial vehicles. Brussels

Committee on Industry Research and Energy to the European Parliament (2008) Opinion of the committee on industry, research and energy for the committee on the environment, public health and food safety on the proposal for a regulation of the European parliament and of the council setting emission performance standards for new pass, vol 0297. Brussels

Committee on the Environment Public Health and Food Safety to the European Parliament (2007) Report on the Community Strategy to reduce CO_2 emissions from passenger cars and light-commercial vehicles. Brussels

Committee on the Environment Public Health and Food Safety to the European Parliament (2008) Report on the proposal for a regulation of the European parliament and of the council setting emission performance standards for new passenger cars as part of the community's integrated approach to reduce CO_2 emissions from light-duty vehicles

Committee on the Environment Public Health and Food Safety to the European Parliament (2012) Report on the proposal for a regulation of the European parliament and of the council amending regulation (EC) No 443/2009 to define the modalities for reaching the 2020 target to reduce CO_2 emissions from new passenger cars

Council of the European Union (2007a) Communication from the commission to the council and the European parliament on the results of the review of the community strategy to reduce CO_2 emissions from passenger cars and light-commercial vehicles—council conclusions. Brussels

Council of the European Union (2007b) Press release 2785th council meeting. Brussels

Council of the European Union (2008a) Press release 2856th council meeting, vol 08. Brussels

Council of the European Union (2008b) Press release 2874th council meeting. Luxemberg

Council of the European Union (2008c) Proposal for a regulation of the European parliament and of the council setting emission performance standards for new passenger cars as part of the community's integrated approach to reduce CO_2 emissions from light-duty vehicles—progress report by the. Brussels

Council of the European Union General Secretariat (1994) Council of the European union: press releases. Luxemberg

Council of the European Union General Secretariat (1996) 1939th council meeting on environment: press release. Brussels

Council of the European Union General Secretariat (1997) 2062nd council meeting on environment: press release. Brussels

Council of the European Union General Secretariat (1998) 2106th council meeting—environment. Luxemberg

Crombez C (2003) The democratic deficit in the European union: much ado about nothing? Eur Union Politics 4(1):101–120

Delreux T (2012) The EU as an Actor in global environmental politics. Environmental policy in the European union: contexts, actors and policy dynamics. Earthscan, London, pp 287–306

EESC (1998) Opinion of the economic and social committee on the 'proposal for a council decision establishing a scheme to monitor the average specific emissions of carbon dioxide from new passenger cars. Retrieved 29 Mar 2012, from http://eur-lex.europa.eu/LexUriServ/LexUriServ.do?uri=CELEX:51998AC1437:EN:NOT

Europa (n.d.) Democratic deficit. Retrieved 30 June 2013, from http://europa.eu/legislation_summaries/glossary/democratic_deficit_en.htm

European Commission (1991) Communication from the commission to the council: a community strategy to limit carbon dioxide emissions and to improve energy efficiency. Brussels

European Commission (2007a) COM(2007) 856 final. Proposal for a regulation of the European parliament and of the council—setting emission performance standards for new passenger cars as part of the community's integrated approach to reduce CO_2 emissions from light-duty vehicles. Brussels

European Commission (2007b) Communication from the commission to the council and the European parliament 6 results of the review of the community strategy to reduce CO_2 emissions from passenger cars and light-commercial vehicles

European Commission (2007c) Press release: commission proposal to limit the CO_2 emissions from cars to help fight climate change, reduce fuel costs and increase European competitiveness commission proposal to limit the CO_2 emissions from cars to help fight climate change, reduce f. Brussels

European Commission (2007d) Public consultation. Brussels

European Commission (2012) COM(2012) 393 final: proposal for a regulation to define the modalities for reaching the 2020 target for reducing CO_2 emissions from new car passenger cars. Brussels

European Parliament (2005) Winning the battle against global climate change (2005/2049(INI))

European Parliament (2007a) MEPs back cuts in cars' CO_2 emissions. Retrieved 31 July 2012, from http://www.europarl.europa.eu/sides/getDoc.do?language=EN&type=IM-PRESS&reference=20071023IPR12110

European Parliament (2007b) Debates—Monday, 22 October 2007—community strategy to reduce CO_2 emissions from passenger cars and light-commercial vehicles, vol 2119. Brussels

European Parliament (2008) Position of the European parliament adopted at first reading on 17 December 2008 with a view to the adoption of regulation (EC) No ../2009 of the European parliament and of the council setting emission performance standards for new passenger cars as part o. Brussels

Fiat (2008) Annual report 2008. Turin

Fiat (2009) Annual report 2009. Turin

Fiat (2010) Annual report 2010. Turin

Friends of the Earth (2007) Consultation on the review of the community strategy to reduce CO_2 emissions from passenger cars and light-commercial vehicles—response from friends of the earth England. Wales and Northern Ireland, Brussels

General Motors (2007) Response from general motors to the public consultation on the implementation of the renewed strategy to reduce CO_2 emissions from passenger cars and light commercial vehicles lower CO_2 alternative fuel solutions in market today. Brussels

Greenpeace (2007) Re: submission to the European commission consultation on the implementation of the renewed strategy to reduce CO_2 emissions from passenger cars and lightcommercial vehicles. Brussels

Greenpeace (2008) Driving climate change: how the car industry lobbying to undermine EU fuel efficiency legislation

Haufler V (2001) Private sector: industry self-regulation in a global economy. Carnegie Endowment for International Peace, Washington DC

ICCT (2011) European vehicle market statistics. Washington DC

Iguchi M (2012) The limitation and possibilities of Japan's proposal for a road transport task force in the Asia-pacific partnership (APP). J Environ Info Sci 40(5):61–68

Iguchi M, Hillman K (2012) The development of fuel economy regulation for passenger cars in Japan. In Nilsson et al (eds) Paving the road to sustainable transport governance and innovation in low-carbon vehicles. Routledge, London, pp 57–69
IEA (2009) CO_2 emissions from fuel combustion highlights—2010 edition. Outlook, Paris
JAMA (2007) JAMA reply to the European commission's consultation on the renewed strategy to reduce CO_2 emissions from passenger cars and light-commercial vehicles. Brussels
Johnson SP, Corcelle G (1989) The Environmental Policy of the European Communities. Graham & Trotman, London
Jordan A, Adelle C (2012) EU environmental policy: contexts, actors and policy dynamics. In: Jordan A, Adelle C (eds) Environmental policy in the European union: contexts, actors and policy dynamics. Routledge, London
KAMA (2007) KAMA's comment to the comissions's consultation on implementation of the renewed strategy to reduce CO_2 emissions from passenger cars and light-commercial vehicles. Brussels
Keay-Bright S (2000) A critical analysis of the voluntary fuel economy agreement, established between the European automobile manufacturers and the European commission, with regard for its capacity to protect the environment. Brussels
Kelemen D (2010) Globalizing European union environmental policy. J Eur Public Policy 17 (3):335–349
Knill C, Lieffernik D (2007) The establishment of EU environmental policy. In: Jordan AJ, Adelle C (eds) The establishment of EU environmental policy, vol Environment. Earthscan, London
Leveque F (1996) The European fabric of environmental regulations. In: Leveque F (ed) Environmental policy in Europe. Edward Elgar, Cheltenham
Lightfoot S, Burchell J (2005) The European union and the world summit on sustainable development: normative power Europe in action? J Common Mark Stud 43(1):75–95
Manners I (2002) Normative power Europe: a contradiction in terms? J Common Mark Stud 40 (2):235–258
Moravsik A (2008) The myth of Europe's democratic deficit. Intereconomics J Eur Public Policy 43:331–340
Official Journal of the European Union (2009) Regulation (EC) No 443/2009 of the European parliament and the council setting emissions performance standards for new passenger cars as part of the community's integrated approach to reduce CO_2 emissions from light-duty vehicles. Brussels
Official Journal of the European Union (2014) Regulation (EC) No 333/2014 of the European parliament and the council amending regulation (EC) NO 443/2009 to define the modalities for reaching the 2020 target to reduce CO_2 emissions from new passenger cars. Brussels
OICA (2007) World motor vehicle production OICA correspondents survery. Paris
OICA (2008) World motor vehicle production OICA correspondents survery. Paris
OICA (2009) World motor vehicle production OICA correspondents survery. Paris
OICA (2010) World motor vehicle production OICA correspondents survery. Paris
PSA Peugeot Citroen (2008) Annual results 2008. Paris
PSA Peugeot Citroen (2010) Annual results 2010. Paris
Steinmo S et al (1992) Structuring politics: historical institutionalism in comparative analysis. Cambridge University Press, Cambridge
T&E (2007a) 2006 Progress report on the car industry's voluntary commitment. Brussels
T&E (2007b) Response to the EU public consultation on the implementation of the renewed strategy to reduce CO_2 emissions from passenger cars and light-commercial vehicles. Brussels
T&E (2009) Reducing CO_2 emissions from new cars : a study of major car manufacturers' progress in 2008. Brussels
T&E (2013) Effect of the German phase-in proposal to the 2020 target. Brussels
The ENDS Report (1992a) Deadlock over CO_2 controls on cars. London
The ENDS Report (1992b) Competing options for EC curbs on CO_2 emissions from cars. London
The ENDS Report (1994) ENDS report on environment council. London, 4 Oct 1994
The ENDS Report (1996) Lengthy EC negotiations ahead on CO_2 strategy for cars. London

The ENDS Report (1997) Agreement on car efficiency heads for the rocks. London
The ENDS Report (1998a) Car makers give ground on fuel efficiency agreement. London
The ENDS Report (1998b) Car makers give ground on fuel efficiency agreement. London
The ENDS Report (1998c) Negotiations with industry on CO_2 emissions from motor vehicles—conclusions. London
The European Commission (n.d.) Road transport: reducing CO_2 emissions from vehicles. Retrieved 30 June 2013, from http://ec.europa.eu/clima/policies/transport/vehicles/index_en.htm
Usui Y (2007) International collective actions for climate change: climate regime, the UNEP and the EU. Keio Law J 8:119–121
VDA (2007) Communication from the commission on reducing CO_2 emissions from passenger cars position of the VDA within the scope of the internet consultation of the European commission political concerns for the German automotive industry. Brussels
Vogel D (1997) Trading up and governing across: transnational governance and environmental protection. J Eur Publ Policy 4(4):556–571
Vogler J, Bretherton C (2006) The European as a protagonist to the United States on climate change. Int Stud Perspect 7(1):1–22
Volkswagen (2008) Volkswagen annual report 2007. Wiesbaden
Volkswagen (2009) Volkswagen annual report 2008. Wiesbaden
Volkswagen (2010) Volkswagen annual report 2009. Wiesbaden
Volkswagen (2011) Volkswagen annual report 2010. Wiesbaden
Weidner H (1995) 25 Years of modern environmental policy in Germany. Treading a well-worn path to the top of the international field. Berlin, pp 1–99
Wurze RKW (2002) Environmental policy-making in Britain. Manchster University Press, Germany and the European Union
WWF (2007) Submission to EC's consultation on the strategy for reducing CO_2 emissions from cars. Brussels

Chapter 4
Construction of Japanese Fuel Economy Regulations for Passenger Cars

Abstract This chapter focuses on how Japanese climate policy for automobiles has been constructed. Japan has the highest fuel economy regulation along with the EU in the world. The question is, did Japan adopt the world's highest fuel economy regulation by same reasons as European case, or are there any distinctive factors? This chapter argues that Japan adopted its stringent fuel economy regulation primarily because of industry competitiveness. The main reason for this rationale to enhance its industry competitiveness by setting stringent regulation is due to its size of the market. Furthermore, this chapter points out that Japan's decision-making process, which is characterized as 'co-regulation' and 'corporatism' between government and the industry, enables to maintain its stringent fuel economy regulations. To do so, this chapter looks at the construction of Japanese fuel economy regulations, by focusing on how foreign markets have been impacted on Japan's fuel economy regulations, as well as how Japan's 'co-regulation' and 'corporatism' have been enhancing the regulations.

Keywords Japan · Car fuel economy regulations · Competitiveness of japanese automobile industry · Corporatism · Co-regulation

4.1 Introduction

Chapter 3 looked at the construction of European fuel economy regulations. It explained both domestic and international political dynamics constructed the EU's highest fuel economy standard among the major automobile manufacturing regions. In particular, it pointed out the EU's effort to internationalize its standard in order to enhance the competitiveness of the industry was fundamental factor that pushed for its stringent standard in the world. Then it concluded that the EU level

Sections 4.2, 4.3 and 4.4 of this chapter are based on Iguchi and Hillman (2012).

M. Iguchi, *Divergence and Convergence of Automobile Fuel Economy Regulations*,
DOI 10.1007/978-3-319-17500-3_4

legislation would influence foreign car manufactures who wish to sell their products in the Europe.

Japan is home to several global car manufactures—Toyota (founded in 1937), Honda (founded in 1948), Nissan (founded in 1933), Suzuki (founded in 1920), Mitsubishi (founded in 1970), and so forth. Among these companies, Nissan (Renault, 1993–), Mazda (Ford, 1979–), Suzuki (Volkswagen, 2009–2011), and Mitsubishi (Daimler-Chrysler, 2000–2005) form capital alliance with foreign companies. In Japan, the car industry is one of the basic industries with about 8 % of the Japanese population being employed in, and relating to the industry (JAMA 2010a). The international landscape in the car sector has changed significantly between 1998 and 2008: the US-dominant market structure that prevailed in 1998 has changed with sales of Toyota ranking at the top in 2008.

Japan's CO_2 emission from fuel combustion is on the rise since 1990.[1] According to IEA (2009), Japan's total CO_2 emission from fuel combustion was 1,092.9 million tons, and transport sector accounts for 20 % (220.1 million tons of CO_2), and 18 % (198.2 million tons of CO_2) of the emission came from Road Transport Sector in 2009.[2] While the energy conservation has been the central rationale for Japanese fuel economy regulation, concerns for climate also played an important role. Especially after the adoptions of the UNFCCC in 1992 and the subsequently the Kyoto Protocol in 1997, Japan has been strengthening its fuel economy regulation due to its concerns on the environment.

This chapter focuses on how Japanese climate policy for automobiles has been constructed. As mentioned in Introduction, Japan has the highest fuel economy regulation along with the EU in the world. The question is, did Japan adopt the world's highest fuel economy regulation by same reasons as European case, or are there any distinctive factors? Previous studies point out following characteristics of Japanese climate policies. Firstly, Japan's domestic interest group politics plays critical part. As Fisher (2004, p. 73) argues, "[t]he regulation of climate change in Japan is a case in point that the government has only taken steps that are approved by industry". In contrary, although the voice of environmental NGOs have been increasingly heard in the government in recent years (Fisher 2004; Tiberghien and Schreurs 2007), their power is not as influential as compared to NGOs in other countries (Foljanty-Jost 2005). Secondly, rivalry between different ministries, especially between the Ministry of Enterprise and Industry (METI) and the Ministry of Environment (MOE) has cultivated Japanese climate politics (Tiberghien and Schreurs 2007). These ministries have their own networks with "interlocking self-interests among bureaucracy, politicians and interest group based on reciprocal political exchange" (Oshitani 2006, p. 68). The common argument is the METI, backed up by Keidanren (the central industry network), has been argued that

[1]In 1990, the total CO_2 emission from fuel combustion in Japan was 1064.4 million tons of CO_2, and the number has increased to 1093.9 million tons of CO_2 in 2009 (IEA 2009, p. 6).

[2]Other sector are as follows (all in millions tons of CO_2): 434.4 from electricity and heat production; 41.3 from other energy industry own use; 238.8 from manufacturing industries and construction; and 158.2 from other sectors including residential sector.

Japan's emission reduction target was proving to be difficult to achieve, especially for the industrial sector (METI 2004). In contrary, the MOE emphasised the support and enhancement of the Kyoto Protocol through agreeing on legally binding numerical targets (MOE 2004), however its voice was not really reflected in Japanese foreign climate policies. The third characteristic relates to Japan's historical and economic ties with the US. It is often pointed out that Japanese security and economic issues has been largely influenced by the US (Iida 1999; Ikenberry and Inoguchi 2003; Inoguchi 1993). Climate change issue is not an exception, since the US is important country in terms of its CO_2 emission which is the second largest in the world after China. Matsumura (2000) observes that Japan has sided with the US on several occasions in the climate change negotiations, although their positions have not always aligned.

These characteristics of Japanese climate policies are not very convincing to explain the stringent Japanese fuel economy regulations. Rather, all these characteristics appear to be stumbling blocks to setting stringent fuel economy regulations: industry-centred climate policies with weak voice from environmental NGOs would undermine stringent fuel economy regulatory standards proposed by government; strong voice of the Economic Ministry over Environmental Ministry would enables to reflect business preferences into the decision-making process; and the ties with the US, one of Japan's biggest trade partner which has stagnant fuel economy regulation for past 20 years, would not motivate Japan to have stringent fuel economy regulation.

Considering all of these characteristics, how and why Japan adopted one of the highest fuel economy standards in the world, along with the EU? This chapter argues that Japan adopted its stringent fuel economy regulation primarily because of industry competitiveness. The main reason for this rationale to enhance its industry competitiveness by setting stringent regulation is due to its size of the market. Furthermore, this chapter points out that Japan's decision-making process, which is characterised as 'co-regulation' and 'corporatism' between government and the industry, enables to maintain its stringent fuel economy regulations. To do so, this chapter looks at the construction of Japanese fuel economy regulations, by focusing on how foreign markets have been impacted on Japan's fuel economy regulations, as well as how Japan's 'co-regulation' and 'corporatism' have been enhancing the regulations.

4.2 The Brief History of the Japanese Fuel Economy Regulation: Mid-1970s—Late 1990s

Japanese fuel economy regulation was triggered by domestic and international events. In the domestic context, the growing concerns about air pollution and accompanying damage to health eventually led to the creation of the Air Pollution Control Law of 1968, which in turn formed the basis of environmental regulations for the automobile industry. An important international event was the first oil crisis in the early 1970s. As a country of low energy sufficiency rate and a high reliant of foreign energy imports, sudden increases in the oil price have been huge impacts on its economy. Hence, the Law Concerning the Rational Use of Energy (Energy Conservation Act) in 1976 was created in order to encourage energy savings in factories, transportation, and buildings. The second oil crisis in the late 1970s resulted in the Amendment of the Energy Conservation Act in 1979, which in turn created the first fuel economy standard for the domestic manufactures' sales of new cars.

Another international event that fostered Japanese fuel economy regulations was the adoption of the UNFCCC in 1992. Since then, Japanese fuel economy regulations have been motivated not only by energy savings but also by concerns about climate change, and led to a new gasoline car fuel economy target for 2000 (Miyoshi and Tanishita 2008). After the Kyoto Protocol was adopted in 1997, the Guideline of Measures to Prevent Global Warming was announced by the newly established Global Warming Prevention Headquarters under the Cabinet Office of the Japanese Government. This guideline largely revised the Energy Conservation Act that introduced the 'top runner method'. This method sets the standard based on the most efficient model—the top runner—in each of a number of weight classes, and obligates the rest of the models to follow the top runner in a given time.

With an introduction of the method, the fuel economy standards were largely strengthened, requiring 22.8 % improvement for gasoline cars as compared to 1995 levels by 2010 (15.1 kilometre per Litre, km/L), and 14.9 % (11.6 km/L) improvement for diesel cars by 2005. Notably, Japanese car manufactures achieved the 2010 target for gasoline cars already by 2007. As a result, a stricter standard, which requires 16.8 km/L (125 g/km) was introduced in 2007 for gasoline cars with the target year 2015. Currently, Japanese 2020 fuel economy standard for passenger cars (105 g/km), which is as stringent as the EU's 2020 target (95 g/km), was set in 2011. These developments are summarized in Table 4.1.

4.3 Impacts from the Regulation Abroad

As one of the major car manufacturing countries, Japan exports great numbers of cars. In doing so, Japanese car companies have invested in considerable research and development (R&D), in order to comply with environmental regulations of importing countries—which are sometimes more stringent than the standards at home—but also for the purpose of gaining the first mover advantage. A classic example of how foreign regulation influenced the strategies of the Japanese car industry is provided by the US Air pollution Act of 1970 [also known as the 'Muskie Act'], which set high standards for every car manufacturer to fulfil. In particular, it encouraged the Japanese car manufacturer Honda in inventing the Compound Vortex Controlled Combustion (CVCC) engine that made Honda the first car company to comply with the regulation (Honda n.d.).

Table 4.1 The milestones of Japanese fuel economy regulations

Year	Policies relating to fuel economy regulations	Target year
1976	Introduction of the Energy Conservation Act	
1979	Introduction of the first fuel economy standard for gasoline cars	1985
1993	Revision of the fuel economy standard for gasoline cars	2000
1996	Introduction of the fuel economy standard for gasoline freight vehicles	2003
1998	Revision of the Energy Conservation Act, with the Top Runner method	
1999	Revision of the fuel economy standards for gasoline and diesel cars, and small freight vehicles	2005 (diesel) 2010 (gasoline)
2003	Introduction of the fuel economy standard for LPG vehicles	2010
2006	Introduction of the fuel economy standard for heavy vehicles (trucks and busses)	2015
2007	Revision of the fuel economy standards for passenger cars and small freight vehicles, and introduction of a new standard for light weight busses	2015
2011	New standard for passenger cars for 2020	2020

Source created by author based on MLIT (2007a)

Up until today, environmental standards in the US and the EU are particularly important in encouraging technological innovation in the Japanese car industry. The US is Japan's biggest car export market with 30.7 % of the total numbers of four-wheel cars exported in 2008, while Europe is the second biggest market with a share of 23.6 % (JAMA n.d.).[3] Although Asia is growing to be among the biggest export markets, its share is not yet as big as those of the US or the EU. Notably, as Table 4.2 clearly indicates, for the major Japanese manufacturers, the absolute sales in the US have been larger than in the domestic market.

Furthermore, both the US and the EU are major bases for local car production of the Japanese car industry. Such production started in the 1980s, when trade frictions between Japan and the US grew with the sharp increase in Japanese car exports. Since then, for more than two decades, the US has been the biggest local producer of Japanese cars as shown in the Table 4.2. For instance, in 2002, 2.7 million cars were produced in the US by the Japanese car industry (35 % of the industry's total overseas production), one million in the EU (14 %), whereas 10 million cars were produced domestically. Although the share of production located to the US then decreased due to a shift in Japanese car industry's local production to Asia, it still accounted for about 25 % of the overseas production in 2008 (31 % in North America). The overseas production in Europe slightly increased its share since year 2000, and accounted for about 15 % in 2008 (JAMA n.d.).

[3]Other shares are as follows: Middle East 14.2 %, Asia 7.8 %, Oceania 6.8 %, Africa 5.2 %, Central Africa 3.8 %, South Africa 3.8 %, and others 0.2 %.

Table 4.2 Motor vehicle production and sale by major Japanese car industry by region, 2007–2010

Year/Company		Toyota	Honda	Nissan
2007	Production	**Japan: 4,226,137 (50 %)** Europe: 645,313 (6 %) US: 1,334,183 (16 %)	**Japan: 1,331,845 (34 %)** Europe: 237,783 (6 %) US: 1,015,462 (25 %)	**Japan: 1,179,080 (34 %)** Europe: 576,632 (17 %) US: 703,662 (20 %)
	Sales	Japan: 27 % Europe: 14 % **US: 35 %**	Japan: 18 % Europe: 9 % **US: 49 %**	Japan: 19 % Europe: 17 % **US: 36 %**
2008	Production	**Japan: 4,012,388 (43 %)** Europe: 561,686 (6 %) US: 1,117,409 (12 %)	**Japan: 1,264,381 (32 %)** Europe: 230,423 (6 %) US: 987,169 (25 %)	**Japan: 1,293,082 (38 %)** Europe: 543,792 (16 %) US: 545,057 (16 %)
	Sales	Japan: 25 % Europe: 14 % **US: 33 %**	Japan: 16 % Europe: 10 % **US: 47 %**	Japan: 18 % Europe: 16 % **US: 25 %**
2009	Production	**Japan: 2,792,274 (39 %)** Europe: 435,261 (6 %) US: 869,564 (12 %)	**Japan: 840,924 (28 %)** Europe: 75,583 (3 %) US: 723,375 (24 %)	**Japan: 894,575 (33 %)** Europe: 390,727 (14 %) US: 372,906 (14 %)
	Sales	Japan: 26 % Europe: 14 % **US: 29 %**	Japan: 16 % Europe: 10 % **US: 43 %**	Japan: 18 % Europe: 15 % **US: 23 %**
2010	Production	**Japan: 3,282,855 (38 %)** Europe: 378,351 (4 %) US: 945,432 (11 %)	**Japan: 992,502 (27 %)** Europe: 139,264 (4 %) US: 954,502 (26 %)	**Japan: 1,133,667 (28 %)** Europe: 528,126 (13 %) US: 511,498 (13 %)
	Sales	**Japan: 30 %** Europe: 12 % US: 29 %	Japan: 19 % Europe: 7 % **US: 38 %**	Japan: 14 % Europe: 16 % **US: 23 %**

Source Toyota (2007, 2008, 2009, 2010), Honda (2007, 2008, 2009, 2010), Nissan (2007, 2008, 2009, 2010), OICA correspondence survey (2007, 2008, 2009, 2010), modified by the author

In the US, where fuel economy regulation at federal level has been stagnant, state-level regulations have been evolved significantly. The best example of stringent state-level regulation of cars is California's low-emission vehicle (LEV) regulation in the 1990s that required all car manufactures to include zero-emission vehicles (ZEVs) as a small percentage of their total sales in the state. Although the regulation was gradually weakened as a result of the industry lobbying activities, it mandated the industry to include 2 % ZEVs in 1998 and 10 % in 2003, including advanced-technology partially zero-emission vehicles such as hybrid cars. According to Yarime et al. (2008) this rigorous regulation pushed technological innovation in hybrid and fuel-cell cars in the Japanese car industry. For example, after the introduction of California's ZEV regulation in 1990, the number of Japanese patent applications on EVs and hybrid cars increased sharply (Yarime et al. 2008).

On the other hand, given the stagnated fuel economy regulation at the Federal level, Japanese car industry has been enjoying its large share in the US market (more description in the next chapter). It argues that Japanese car industry create about 1.36 million private sector US jobs, thus contributing to the US economy, and hence, the free-trade agreement under the Trans-Pacific Partnership (TPP) would contribute to the US economy and provide more opportunities for creating employments (JAMA 2013a, b).

In Europe, which has been leaning the worlds' fuel economy regulations, a voluntary target was agreed between JAMA and the European Commission in 1998, aiming for the target of 140 g/km by 2009. Given that the average fuel economy of Japanese car industry in 2000 was only 169 g/km, this was a stringent target for the industry (ICCT 2014). As a result, this target may have worked as one of the driving forces that pushed improvement in the Japanese industry, which improved up to 129 g/km in 2009 (ICCT 2014).

Nevertheless, it is apparent that the European market is of great importance to the Japanese auto manufactures. According to JAMA-Europe's report, titles as 'Common Challenge, Common Future: Japanese Auto Manufactures Contribute to the Competitiveness of Europe's Motor Industry', it emphasises that:

> The global urgency of addressing climate change has prompted vital discussions on government policies to promote eco-friendly vehicles and on strengthening the competitiveness of the automotive industry in the transition to decarbonized road transport. Despite of the challenges posed by the global economic slowdown since autumn 2008, Japanese automobile manufactures are contributing to the economy and employment in the European Union through their production, research-and-development, distribution and other activities in the EU, and contributing to the shift to low-carbon transport through their supply of eco-friendly vehicles (JAMA 2010b, p. 2).

Also important is the discussions associating with the Free Trade Agreement (FTA) between Japan and the EU that started at the EU-Japan Summit on May 2009 that engaged into a formal negotiation process since March 2013 (Council of the European Union 2011; European Commission 2013). Since Japanese auto industry have a competence in the European market, JAMA supports the bilateral trade agreement, and even "advocates the early conclusion" of the FTA between the EU and Japan (JAMA 2010b, 2011, 2012, 2013c, 2014), while European automobile manufactures are generally reluctant to such agreements. ACEA argues for following two expectations: first, European cars should be sold in Japanese market without further testing or modification; and second, an opportunity should be given to European small cars to compete with Japanese cars of same size on equal terms, since the EU exports to Japanese market represent 3.5 % of units whilst Japanese car represents 18 % share of EU imports (ACEA n.d.).

4.4 Critical Juncture: 2007

The critical juncture to the Japanese fuel economy regulations was brought in 2007, when Japanese car manufactures achieved the 2010 target already by 2007. As a result, new target for 2015 was introduced by the revision of Energy Conservation Act in 2007, marking 125 g/km which was as stringent as Europe's proposed 120 g/km target of that time. The fundamental rationale of the regulation is due to the following two reasons. First, Japan experienced the increase of CO_2 emissions up to 8.1 % compared to 1990 levels in 2005, thus emission reduction from the road transport sector was critical to achieve the commitment made under the Kyoto Protocol. Second, facing increasing demands of energy use in newly emerging economies and rising price of oil, Japan introduced the 'New National Energy Strategy' (adopted in May 2006) and 'outline of Economic growth Strategy' (adopted in July 2006) to set 30 % improvement of energy efficiency by 2030 and to reduce oil dependence in transport sector by 80 % (METI and MLIT 2007). The question is then, despite the fact that Japanese auto industry had already satisfied 2010 target in 2007, why they did not opposed to the further regulation, and attempted to delay the regulation process? This section argues that the answer lies in the following two characteristics of Japanese fuel economy regulations.

The first characteristics of Japanese fuel economy regulation is that it regulate cars' fuel efficiency based on different weight segmentations, with the rationale to make competition fair in each category. It introduces fuel economy standards according to nine vehicle weight classes. This can be compared with the corporate average fuel economy (CAFE) regulations in the US, which addresses the average fuel economy of each manufacturer's sales of passenger cars and light trucks, respectively. The implication would be that while American car manufacturers can follow a strategy of increasing the sales of light weight vehicles in order to fulfil the requirement, Japanese car manufacturers cannot follow this strategy since fuel regulations are segmented by vehicle weight.

One of the results of the Japanese regulation is that it pushed the Japanese car industry to be one of the most fuel efficient industries among the major car manufacturing countries. Due to the large share of domestic brands—increasing from 65 to 87 % between 2002 and 2008 (JADA n.d.; JAIA n.d.)—this is also reflected in the fleet average CO_2 emissions of passenger cars sold in the three major car manufacturing regions in that period. Table 4.3 shows the comparison of actual fleet

Table 4.3 Comparison of actual fleet average CO_2 emission of passenger cars sold in each region 2002–2008

Year	US	Canada	EU	Japan	China	South Korea
2002	219	201	167	157	213	–
2003	216	199	166	156	–	232
2004	217	196	163	154	–	218
2005	212	193	162	153	–	214
2006	214	196	161	149	188	213
2007	207	187	159	147	–	207
2008	205	184	154	141	185	198
2009	194	176	146	129	–	184
2010	188	–	140	128	180	175
2011	190	–	136	119	176	167
2012	178	–	132	110	172	–
2013	175	–	127	–	–	–

Source created by author based on ICCT (2014)

average CO_2 emission of passenger cars sold in each region from 2002 to 2008. They show that Japan, along with the EU, is leading the fuel economy of passenger cars in the world.

Furthermore, in order to promote Japan as a base also for future car production, METI proposed in 2010 a sustainable technology road map for the Japanese car industry. According to the road map, it aims to reduce the share of conventional internal combustion engine (ICE) cars, while increasing the share of next generation cars, such as hybrid, plug-in hybrid, electric, fuel cell, biofuel and 'clean diesel' cars up to 70 % by 2030. Table 4.4 shows the Japanese technology roadmap for the next generation vehicles.

While it sets the aim to reduce the sales of conventional ICE cars on the one hand, the goal for 2020 point to only 50–80 % reduction. This is mainly because high efficient ICE cars are essential for the Japanese car industry to strengthen its competitiveness on the global market, where ICE cars are expected to be the mainstream type of car in newly emerging economies (METI 2010). Accordingly, METI's car technology roadmap suggests that both next generation cars, headed by hybrid electric vehicles (HEVs) and plug-in hybrids, as well as high efficient ICE are crucial technologies for the Japanese car industry in the next 20 years.

The second characteristics of Japanese car fuel economy standards are formulated based on government –industry consultations. According to Schreurs (2003) and Mikler (2010), Japanese environmental policy heavily relies on regulations, but of a 'Japanese form', which is long rooted in close industry–government consultations: government first shows 'administrative guidance' (*gyosei shido*)[4] to the

[4]In Article 2 of the Administrative Procedure Act from in 1993, administrative guidance is defined as "guidance, recommendations, advice, or other acts by which an Administrative Organ may seek […] certain action or inaction on the part of specified persons in order to realize administrative aims" (Cabinet Secretariat n.d., p.3).

Table 4.4 Japanese technology road map for the next generation cars

	2010	2015	2020	2030
Reducing share of conventional cars			50%~80%	30~50%
		High efficient internal combustion engine →		
Increasing share of next generation cars			20%~50%	50%~70%
		Current hybrid →	New hybrid	
		Commuter EV →	High performance EV	
		Plug-in hybrid →	High performance plug-in Hybrid	
			Fuel cell →	Next generation fuel cell
	Bio-gasoline →		Cellulose biofuel →	
	Clean diesel →			

Source created by author based on METI (2010) and Nikkei (2009)

industry, by giving recommendations and advices in order to achieve a certain goal. Based on this consultation, it is commonplace that the industry follows the government's guidance through some form of self-regulation. After changes in industry can be observed, government pushes for stricter regulation. This Japanese form of regulation can be grasped by the concept of 'co-regulation'. The concept is more commonly used by practitioners than in governance literature, and there seems to be no conventional definition of the term. Nevertheless, it has close linkage to what Schmitter (1974) calls 'societal corporatism': the state and corporations negotiate policy by consultations, although the members of corporations do not have significant involvement in the process. Accordingly, the concept can be defined as the mix of government regulation and self-regulation based on collaboration between the government and the industry (Europa 2004; OFTEL 2000) or, 'regulated self-regulation' (Schulz and Held 2004).

The concept of the corporatism suggests that state decision-making is based on 'tripartism' between state, labour and corporations as social partners to create particular policies through cooperation, consultation, negotiation and compromise (Wiarda 1996). For instance, corporations ask for their preferred policy to government and stable provision of workforce to labours; labours in turn ask for stable employment rate to corporations and more welfare benefits to the government. The degree of corporatism varies according to the context of each country (Cameron 1984). Japanese corporatism is described as 'corporatism without labour', in which labour unions does not participate in the decisions-making process (Pempel and

Tsunekawa 1979). This suggests that corporations are likely to have more influence over the state decision-making in the absence of strong labour union, and hence allows to formulate policies for the sake of enhancing international competitiveness rather than to the wealfare of the workers.

In the case of Japanese fuel economy regulation, the concept of 'co-regulation' fits well in describing the process. The ministries of Land, Infrastructure, and Transport (MLIT), and Economy, Trade and Industry (METI) are responsible for the regulations.[5] In relation to the 2015 regulation these two Ministries held a series of closed meetings in 2004–2006, and invited the Japanese car industry to participate (Iguchi 2009; MLIT 2007b). The industry took part in the decision making process through the JAMA, which is the central industry network for the Japanese car industry.[6] In the case of the 2015 regulation, after an agreement was reached in these meetings, it was passed on to the Council for examination.[7] The Council consists of several meetings, including the transport policy council, and the industry structure council, and its members are generally chosen from the Japanese academia.[8] Their role in this process is to discuss the appropriateness and feasibility of the new standard, by examining the Japanese car industry's technological potential. After being examined by the Council, the agreement was published as an 'intermediate report', which was open for public consultation (MLIT 2006). As there was not much dissenting opinion, it became the new fuel economy standard under the Energy Conservation Act. Figure 4.1 captures these decision-making procedures of Japanese fuel economy regulations.

What is notable about Japanese fuel economy decision-making is that environmental NGOs play hardly any role in this process. This makes Japanese fuel economy regulation a very peculiar case, because the stringent fuel economy standards are oftentimes proposed by civil society. In the context of the European fuel economy regulation process, the environmental NGOs such as T&E and Greenpeace were instrumental to lead the Commission's proposal of legalizing 120 g/km target by disclosing the incompliance of the industry to the voluntary agreement of 1998. Furthermore, they excreted influences over the Environmental Committee of the European Parliament by pushing for the 95 g/km target by 2020.

In the case of the US fuel economy regulation, too, environmental NGOs such as Sierra Club played a crucial role to educate the public and media about the benefits of increasing fuel economy, work with allies on the Capitol Hill to support for the

[5]MLIT is the official governmental body responsible for transport matters, and METI, which is originally responsible for economic activities, is responsible for the fuel economy regulations due to its authority conferred by the Energy Conservation Act.

[6]Keidanren (the central Japanese business network) does not have a role in the fuel economy regulation in this process.

[7]Based on interview with Mr. Akihiko Hoshi, Deputy Director, Japanese Ministry of Infrastructure, Land, Transport and tourism (MLIT). Interviewed at Tokyo, Japan (4th February 2010).

[8]The Council was chaired by a Professor of Fukui University of Technology, and vice chair was a professor from Waseda University. There are eleven committees, and six members are chosen from university or research institute, and others are chosen from industry networks, such as JAMA.

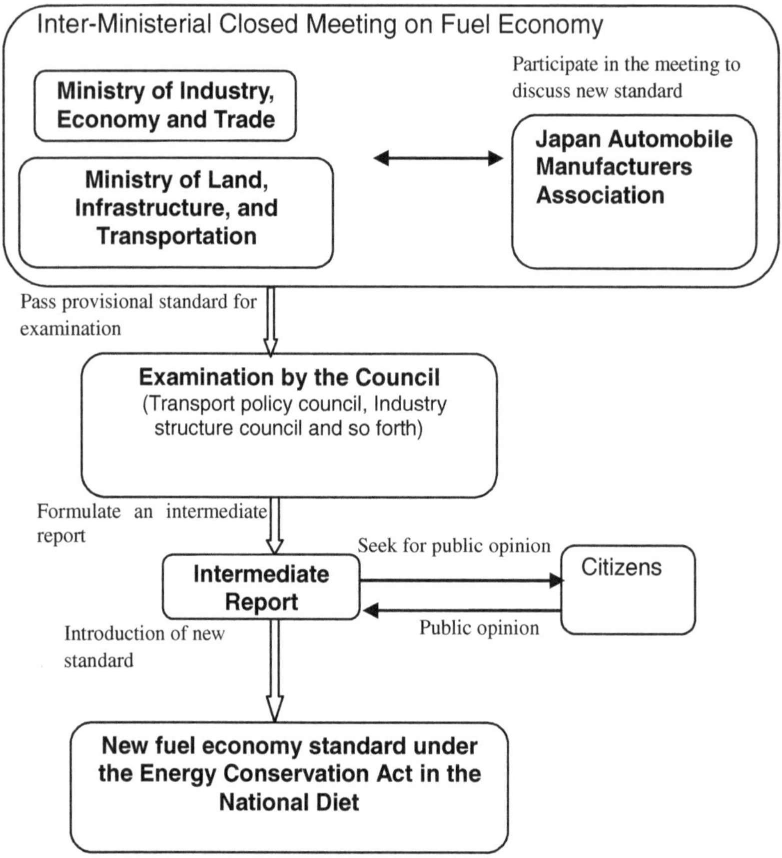

Fig. 4.1 Decision-making process for Japanese fuel economy regulations. *Source* Iguchi and Hillman (2012)

stronger standards, and expressing their opinion to the National Highway Traffic Safety Administration (NHTSA) and the Whitehouse (greater details in next chapter).

In contrast, Japanese environmental NGOs do not have roles in the policy process associating with its fuel economy regulations. Nor there is no policy coordination among Japanese NGOs and its counter parts to influence the process.[9] Furthermore, the public comments, which are not only open to environmental NGOs but also the anyone including the industry, does not hold strong influence to the outcome of the regulation.

[9]Based on interviews conducted to various NGOs in Europe and the US.

4.5 Target for Beyond 2015

It was notable that the actual average fuel economy of Japanese car manufactures in 2010 reached 128 g/km—just 3 g/km above the 2015 target—and achieved 119 g/km in 2011 and hence they had satisfied the 2015 target already (ICCT 2014). This is a significant improvement as compared to the European and American manufactures, both of their actual average fuel economy standards in 2011 were 136 and 190 g/km, respectively.

The consultations to set 2020 target was started in June 2010 and concluded on October 2011 (METI and MLIT 2011a, b). Seven consultations were held in total, and the discussions were focused on the target year, targeted vehicle type, applicability of the top runner method and the method of regulations as well as hearing to automobile manufactures. Just like 2015 target, both METI and MLIT led the regulatory process: an intermediate report was published on August 2011, and after the report was opened to the public, the final agreement was reached on March 2013.

The rationale of the new regulation, as appears in the official document, is primarily to reduce CO_2 emissions from transport sector which accounts for 20 % of the country's total emissions: despite Japanese CO_2 emissions decreased 4.1 % compared to 1990 because of economic stagnation caused by final crisis in 2008 and improvements in energy efficiencies, CO_2 emissions from transport sector increased 5.8 % compared to 1990 (METI and MLIT 2011a, b). Another factors that may influence the regulatory process would be the progressive regulations in Europe and the US. At that time, the European 95 g/km target was clearly mandated in the final document text, and also the possibility of 99 g/km target was announced by the President Obama in July.

The target year of the regulation was set in consideration of production development cycles of the automobile manufactures, which is assumed as the five years. Therefore the target year of 2020, five years after the 2015 fuel economy regulation, was chosen (METI and MLIT 2011a, b). It follows, although one of the rationales of the regulation is climate-driven, the regulation itself was not formulated in accordance with the Japan's mid-term emission reduction pledge to the UNFCCC. Rather, although competitiveness issues are not formally written as the reasons of raising fuel economy regulation, the competitiveness concerns were at the basis of the Japanese target for 2020.

It is also notable that 2020 regulation employs the CAFE as the basis of the regulation, which is already introduced in the Europe and the US.[10] According to METI and MLIT (2011b), while the weight-based regulations encourage fair competition among industry within different weight categories, the CAFE regulations allows each manufactures to choose to invest on particular low-carbon

[10]Although both the EU and the US employs the CAFE regulation, there is a difference in the target setting. While the European CAFE is based on the average vehicle weight, the US regulation is based on the size of the vehicles. Japanese CAFE regulation employs weight-based approach.

vehicles, so that it gives flexibility to corporate strategies. As a result, it promotes diversification and sophistication of low-carbon vehicles technologies, which in turn contributes to the competitiveness of Japanese automobile manufactures. Furthermore, since achievement of the regulatory standard is evaluated based on performances of each manufactures under the CAFE regulation, corporate brand image will be undermined in the case of incompliance.

4.6 Summary

This chapter looked at how Japanese climate policy for car CO_2 emissions has been constructed. Table 4.5 summarizes the characteristics of Japanese fuel economy regulation.

In Japan, motive for fuel economy regulations began in reaction to the oil crisis in 1970s, but later it was strengthened accompanying with the adoption of the UNFCCC in 1992. In the following years, a new fuel economy standard for gasoline cars was introduced for year 2000. After the Kyoto Protocol was adopted in 1997, the Guideline of Measures to Prevent Global Warming was announced by the newly established Global Warming Prevention Headquarters under the Cabinet Office of the Japanese Government. This guideline largely revised the Energy Conservation Act and hence significantly raised Japanese fuel economy regulations, with the introduction of the top runner method.

Unlike the European case, Japanese automobile industry did not really acted as laagered. In fact, one of the main reasons why Japan introduced stringent 2015 fuel economy standard is that Japanese car manufactures achieved the 2010 target for gasoline cars already by 2007. Thus, in Japan, incentives to catch up with stringent fuel economy regulation in the foreign markets as well as co-decision making process which is based on government-industry consultations enabled Japan to set as stringent regulations as the European standards. Based on this consultation, it is

Table 4.5 The characteristics of Japanese fuel economy regulation

Key factors	Description
Motive	Mix of energy-saving and to limit growing CO_2 emissions from the road transport sector
Competitiveness issues	Targets are formulated in order to enhance industry competitiveness, especially in the US and the EU
Decision-making process	Co-regulation between the government and the industry
Business	Industry join decision-making procedure through the industry network
NGOs	Hardly played any role
Critical juncture	In 2007, when industry achieved 2010 target; and hence allowed 2015 target as stringent as the EU's

Source created by author

commonplace that the industry follows the government's guidance through some form of self-regulation. After changes in industry can be observed, government pushes for stricter regulation. It is also notable that environmental NGOs play hardly any role in this process in contrary to the European case. This implies that despite the absence of encouragements from environmental NGOs, Japan set stringent fuel economy regulations.

A critical juncture that directed Japan towards regulatory convergence happened in 2007, when the '2015 regulation' further strengthened the Japanese regulation, which is as stringent as the European target. One of the main reasons of this significant improvement of the regulatory standard is that Japanese car manufactures achieved the 2010 target for gasoline cars already by 2007. As a result, a stricter standard, which requires 16.8 km/L (125 g/km) was introduced in 2007 for gasoline cars with the target year 2015. This target is as stringent as the European target for 2015, and hence it is natural to assume that Japan tried to catch up with the European standard in order to enhance its industry competitiveness.

Another important factor that cultivated stringent fuel economy regulation lies in its decision-making process. In Japan, the development of fuel economy regulation was achieved through 'co-regulation', where government, industry, and academia participated in the standard setting process. In this process, the MLIT and the METI held a series of closed meetings, and invited the Japanese car industry to participate. The industry took part in the decision making process through the central industry network for the Japanese car industry, JAMA. After an agreement was reached in these meetings, it was passed on to the Council for examination. Their role in this process is to discuss the appropriateness and feasibility of the new standard, by examining the Japanese car industry's technological potential. After being examined by the Council, the agreement was published as an 'intermediate report', which was open for public consultation. As there was not much dissenting opinion, it became the new fuel economy standard under the Energy Conservation Act. These series of decision-making procedures enabled incremental improvements of the Japanese fuel economy regulations, in contrast to the EU regulation that took 13 years to legalize its fuel economy regulations.

The most important factor that cultivated Japanese fuel economy regulation towards the regulatory convergence is the competitiveness issues. It was argued that Japanese stringent fuel economy regulation was primarily motivated by stringent regulations in Europe and the US. This is due to the relatively small size of the Japanese car market, and Japanese car industry gain profits from sales in the EU and the US. Table 4.6 compares the timing of fuel economy regulations between Europe, Japan and the US.

As pointed out, Japan exports great numbers of cars. The US is Japan's biggest car export market with 30.7 % of the total numbers of four-wheel cars exported in 2008, while Europe is the second biggest market with a share of 23.6 %. Although Asia is growing to be among the biggest export markets, its share is not yet as big as those of the US or the EU.

Table 4.6 A comparison of the timing of fuel economy regulations between Europe, Japan and the US

Year/event	Milestones of EU fuel economy regulations	Milestones of Japanese fuel economy regulations	Milestones of US fuel economy regulations
1990s	• 1995: the commission proposed '120 g/km target by 2005' • 1998: A voluntary agreement of '140 g/km target by 2008'	• 1998: introduction of 2010 target which required 22.8 % improvement compared to 1995 level	• 1990: California's LEV and ZEV regulations
2000s	• 2007: the commission proposed mandate reduction of '120 g/km (130 g/km) by 2012' and possibility to research towards '95 g/km by 2020'	• 2007: introduction of '125 g/km by 2015' target	• 2002: California passed 'Pavely Law' that set '323 g/km by 2009'; '205 g/km by 2012'; and '172 g/km by 2016' • 2007: Supreme Court decision of 'CO_2 as air pollutant' • 2007: Energy Independence and Security Act mandated '172 g/km by 2020' • 2009: President Obama proposed '172 g/km by 2016'
2010s	• 2011: the commission is reviewing to propose mandate target of '95 g/km by 2020'	• 2010: Started to set '105 g/km by 2020' target • 2011: introduction of '105 g/km by 2020' target	• 2011: President Obama proposed maximum '99 g/km by 2025 target' • 2011: '103 g/km by 2025' agreed

Source created by author

It is argued that stringent environmental regulation for cars in these markets have been encouraging Japanese technologogical innovation. For instance, examples highlighted that California's progressive climate related regulations for cars pushed technological innovations in hybrid electric cars among Japanese car manufactures; and evidence suggests that the actual fuel efficiency of Japanese car manufactures significantly improved after the voluntary targerted being agreed between JAMA and the European Commission, aiming for the target of 140 g/km by 2009.

If we compare timings with the European fuel economy regulation, it is obvious that Japanese regulations are introduced as almost the same timing the European new regulation has proposed. For example, when the discussion of legalization of European 120 g/km target emerged in 2007 and adopted in 2009, which is generally considered as the political target rather than the target based on technological potentials, the Japanese target for 2015 (125 g/km) is also adopted in 2009. Furthermore, as the European 95 g/km by 2020 target was also embedded in the 2009 legislation, Japan began to set 105 g/km by 2020 target right after 2010. Thus, catching up to the strident standards in these exporting markets was essential for the industry to stay competitive in these regions. Therefore, the regulations of overseas are also important factor to analyse Japanese fuel economy regulations.

It follows, it was clear that Japanese regulatory convergence was emerge out of efforts of automobile industry to stay competitive in foreign markets, especially in the EU, where environmental standards are more stringent than home country. The automotive industry, in turn, actively supported to raise the regulatory standard of home country by achieving the 2010 target for gasoline cars by 2007, and accepted stringent government regulation of home country in order to gain a competitive advantage against competitors.

References

ACEA (n.d.) Japan. http://www.acea.be/industry-topics/tag/category/japan. Accessed 3rd Sep 2014

Cabinet Secretariat (n.d.) Administrative Procedure Act. http://www.cas.go.jp/jp/seisaku/hourei/data/APA_2.pdf . Accessed 29 Mar 2015

Cameron D (1984) Social democracy, corporatism, labour quiescence and the representation of economic interests in advanced capitalist society. In: Goldthorpe J (ed) Order and conflict in contemporary capitalism. Oxford University Press, Oxford

Council of the European Union (2011) 20th EU-Japan summit brussels, 28 May 2011 joint press statement. Council of the European Union, Brussels

Europa (2004) Mr Erkki Liikanen member of the European commission, responsible for enterprise and the information society "regulating for competitiveness and growth the regulators experience" IBEC EU presidency conference Dublin, 19. http://europa.eu/rapid/pressReleasesAction.do?reference=SPEECH/04/182&format=HTML&aged=0&language=EN&guiLanguage=en. Accessed 29 March 2012

European Commission (2013) Joint statement by the president of the European commission, Jose Manuel Barroso, the president of the European council, Herman Van Rompuy, and the Prime Minister of Japan, Shinzo Abe. European Commission, Brussels

Fisher D (2004) National governance and the global climate change regime. Rowman & Littlefield, Lanham

Foljanty-Jost G (2005) NGOs in environmental networks in Germany and Japan: the question of power and influence. Soc Sci Jpn J 8(1):103–117

Honda (n.d.) CVCC. http://www.honda.co.jp/environment/glossary/07.html. Accessed 29 March 2012

Honda (2007) Annual report 2007. Tokyo

Honda (2008) Annual report 2008. Tokyo

Honda (2009) Annual report 2009. Tokyo

Honda (2010) Annual report 2010. Tokyo

ICCT (2014) Global passenger car fuel economy and/or greenhouse gas emissions standards—May 2014 update. Washington DC

IEA (2009) CO_2 emissions from fuel combustion highlights—2010 edition. Outlook, Paris

Iguchi M (2009) The agency within the states: the automobile industries in Japan, the Europe, and the U.S. over the architecture of the post-2012 institutional framework. In 2009 Amsterdam Conference on the Human Dimensions of Global Environmental Change

Iguchi M, Hillman K (2012) The development of fuel economy regulation for passenger cars in Japan. In: Nilsson et al. (ed) Paving the road to sustainable transport governance and innovation in low-carbon vehicles. Routledge, London, pp. 57–69

Iida K (1999) International monetary cooperation among the United States, Japan, and Germany. Springer, Berlin

Ikenberry GJ, Inoguchi T (2003) Reinventing the alliance: U.S.-Japan security partnership in an era of change. Palgrave Macmillan, Basingstoke

Inoguchi T (1993) Japan's foreign policy in an era of global change. Continuum International Publishing, London

JADA (n.d.) The number of sales of passenger vehicles in Japan. http://www.jada.or.jp/contents/data/type/index02.php. Accessed 29 March 2012

JAIA (n.d.) The number of sales of imported passenger vehicles in Japan. http://www.jaia-jp.org/j/stat/nc/. Accessed 29 March 2012

JAMA (n.d.) The number of overseas production by Japanese auto industry. http://www.jama.or.jp/stats/foreign_prdct/index.html. Accessed 29 March 2012

JAMA (2009) Common challenge, common future: Japanese auto manufactures contribute to the competitiveness of Europe's motor industry. JAMA EU, Brussels

JAMA (2010a) The percentage of the workforce in Japanese automobile industry. http://www.jama.or.jp/industry/industry/industry_1g1.html. Accessed 30 Aug 2012

JAMA (2010b) Common challenge, common future: Japanese auto manufactures contribute to the competitiveness of Europe's motor industry. JAMA EU, Brussels

JAMA (2011) Common challenge, common future: Japanese auto manufactures contribute to the competitiveness of Europe's motor industry. JAMA EU, Brussels

JAMA (2012) common challenge, common future: Japanese auto manufactures contribute to the competitiveness of Europe's motor industry. JAMA EU, Brussels

JAMA (2013a) Japanese-brand vehicles in America: free trade, greater growth, more jobs. JAMA US, Washington DC

JAMA (2013b) Japanese-Brand vehicles are integral to US economy. http://jama.org/media-center/japanese-brand-vehicles-integral-u-s-economy/. Accessed 3 Sep 2014

JAMA (2013c) Common challenge, common future: Japanese auto manufactures contribute to the competitiveness of Europe's motor industry. JAMA EU, Brussels

JAMA (2014) Common challenge, common future: Japanese auto manufactures contribute to the competitiveness of Europe's motor industry. JAMA EU, Brussels

Matsumura H (2000) Japan and the Kyoto protocol, conditions for ratification. The Royal Institute of International Affairs, London

METI (2004) An interim report by special committee on a future framework for addressing climate change under industrial structure council. Tokyo

METI (2010) Japanese technology road map for the next generation automobile. Tokyo

METI and MLIT (2007) Final report of new fuel economy regulation for passenger cars. Tokyo

METI and MLIT (2011a) Intermediate report of new fuel economy regulation for passenger cars. Tokyo

METI and MLIT (2011b) Final report of new fuel economy regulation for passenger cars. Tokyo

Mikler J (2010) Business as usual or apocalypse now? regulating the car industry's carbon emissions. In: third biennial conference of the european consortium on political research standing group on regulatory governance, Dublin

Miyoshi H, Tanishita M (2008) Jidousha no Gijyutsukakushin to Keizai Kousei (technological innovation in the automobile industry and economic welfare). Hakutoshobo, Tokyo

MLIT (2006) Intermediate report and public opinion for the new fuel economy regulation for passenger cars. http://www.mlit.go.jp/kisha/kisha06/09/091215_.html. Accessed 04 March 2013

MLIT (2007a) Japan's fuel economy regulations. http://www.mlit.go.jp/jidosha/sesaku/environment/ondan/ondan.htm. Accessed 30 March 2012

MLIT (2007b) Regarding new fuel economy regulation for passenger cars. http://www.mlit.go.jp/kisha/kisha07/09/090702_.html. Accessed 04 March 2013

MOE (2004) Climate regime beyond 2012: key perspectives. Interim Report by the Sub-Committee for International Climate Change Strategy, Tokyo

Nikkei (2009) Jidousha Shinseiki: Shousha no Joken (The New era of automobile industry). Nikkei Publishing Inc, Tokyo
Nissan (2007) Annual report 2007. Tokyo
Nissan (2008) Annual report 2008. Tokyo
Nissan (2009) Annual report 2009. Tokyo
Nissan (2010) Annual report 2010. Tokyo
OFTEL (2000) Encouraging self- and co-regulation in telecoms to benefit consumers. London
OICA (2007) World motor vehicle production OICA correspondents survery. Paris
OICA (2008) World motor vehicle production OICA correspondents survery. Paris
OICA (2009) World motor vehicle production OICA correspondents survery. Paris
OICA (2010) World motor vehicle production OICA correspondents survery. Paris
Oshitani S (2006) Global warming policy in Japan and Britain, interactions between institutions and issue characteristics. Manchester University Press, Manchester
Pempel T, Tsunekawa K (1979) Corporatism without labor? The Japanese anomaly. In: Lehmbruch G, Schmitter PC (eds) Trends toward corporatist intermediation. Sage, Beverly Hills
Schmitter PC (1974) Still the century of corporatism. The Rev Politics 36(1):85–131
Schreurs MA (2003) Environmental politics in Japan, Germany and the United States. Cambridge University Press, Cambridge
Schulz W, Held T (2004) Regulated self-regulation as a form of modern government: an analysis of case studies from media and telecommunication law. Eastleigh
Tiberghien Y, Schreurs MA (2007) High noon in Japan: embedded symbolism and post-2001 kyoto protocol politics. Glob Environ Polit 7(4):70–91
Toyota (2007) Annual report 2007. Tokyo
Toyota (2008) Annual report 2008. Tokyo
Toyota (2009) Annual report 2009. Tokyo
Toyota (2010) Annual report 2010. Tokyo
Wardsauto (2014) U.S. Vehicle sales market share by company, 1961–2013. http://wardsauto.com/keydata/historical/UsaSa28summary. Accessed 12 Aug 2014
Wiarda H (1996) Corporatism and comparative politics. M.E. Sharpe, New York
Yarime M et al. (2008) The strategies of Japanese auto industry in developing hybrid and fuel-cell vehicles. In: BLKM, Grant (eds) Making choice about hydrogen: transport issues for developing countries. United Nations University Press, Tokyo

Chapter 5
Construction of the US Fuel Economy Regulations for Passenger Cars

Abstract This chapter focuses on the construction of the US fuel economy regulations. In comparison to the stringent European fuel economy regulations introduced in the early 1990s, the US fuel economy regulatory standards, originally introduced in 1978, are the lowest among the major automobile manufacturing nations. Indeed, US fuel economy regulations have been stagnant for more than 20 years (since the mid-1980s). The critical change was brought about only recently, under the George W. Bush Administration (Republican, 2001–2007), with an introduction of the 2007 Energy Security and Independence Act, which raised the mandatory fuel economy regulation. Under the Obama Administration (Democrat, 2007—present), US fuel economy regulatory standards are now catching up with Japanese and European standards. Consequently, the question is why, despite the US being the world's first country to introduce fuel economy regulations, has US fuel economy regulation been stagnant for more than 20 years? What political dynamics pushed the former Bush Administration, which had withdrawn from the Kyoto Protocol negotiation in 2001, to improve the fuel economy regulation standard? Why are the recent US fuel economy regulations now converging with the Japanese and European standards? This chapter attempts to answer these questions and sketch the logic of the regulatory convergence of the fuel economy among major automobile manufacturing nations.

Keywords The United States · Corporate average fuel economy (CAFE) regulations · Competitiveness of US automobile industry · California effect · Multiple fuel economy regulations

5.1 Introduction

This chapter focuses on the construction of the US fuel economy regulations. In comparison to the stringent European fuel economy regulations introduced in the early 1990s, the US fuel economy regulatory standards, originally introduced in 1978, are the lowest among the major automobile manufacturing nations. Indeed,

M. Iguchi, *Divergence and Convergence of Automobile Fuel Economy Regulations*,
DOI 10.1007/978-3-319-17500-3_5

US fuel economy regulations have been stagnant for more than 20 years (since the mid-1980s). The critical change was brought about only recently, under the George W. Bush Administration (Republican, 2001–2007), with an introduction of the 2007 Energy Security and Independence Act, which raised the mandatory fuel economy regulation. Under the Obama Administration (Democrat, 2007—present), US fuel economy regulatory standards are now catching up with Japanese and European standards (ICCT 2014).

Consequently, the question is why, despite the US being the world's first country to introduce fuel economy regulations, has US fuel economy regulation been stagnant for more than 20 years? What political dynamics pushed the former Bush Administration, which had withdrawn from the Kyoto Protocol negotiation in 2001, to improve the fuel economy regulation standard? Why are the recent US fuel economy regulations now converging with the Japanese and European standards? This chapter attempts to answer these questions and sketch the logic of the regulatory convergence of the fuel economy among major automobile manufacturing nations. The next section looks at the construction of US fuel economy regulations since the 1980s.

There is no doubt that the automobile industry has been the foundation of the US economy; more than one million Americans are employed in the automobile industry, including equipment and parts. The so-called 'Big Three' or 'Detroit Three' companies (General Motors, Ford and Chrysler) have created more than 600,000 jobs since the 1980s. Approximately one quarter of all Americans are employed in the automobile industry (Cooney and Yacobucci 2005). With its mode of mass-production and consumption, commonly known as 'Fordism', the foundation of the automobile industry's importance to the US economy was constructed during the post-second World War period of the 1950s (Gartman 2004). Because of much less availability of public transportation and a larger country in terms of land and mass, in the US, personal mobility, such as that available with cars, has considerably more value than it does in other countries. Accompanying the economic growth of the industry, CO_2 emissions from fuel combustion have increased significantly.[1] According to the IEA (2009a, p. 69), the total CO_2 emission from fuel combustion in 2009 was 5195 million tons: the transport sector was responsible for about 32 % (1614.3 million tons of CO_2) and 27 % (1402.8 million tons of CO_2) of the total CO_2 emissions in the US came from the road transport sector.[2] This section looks at the development of the US fuel economy regulation. It is divided into the following periods: formative years during the 1970s; stagnation from the mid-1980s to the mid-2000; the critical juncture in 2007; and towards regulatory convergence from 2009 onwards.

---

[1]For example, in 1990, the total CO_2 emission from fuel combustion in the U.S. was 4868.7 million tonnes of CO_2, and the number has increased quite dramatically to 5,195 million tonnes of CO_2 in 2009 (IEA 2009b, p. 46).

[2]Other sector are as follows (all in millions tonnes of CO_2): 2190.2 from electricity and heat production; 257.2 from other energy industry own use; 544.4 from manufacturing industries and construction; and 588.8 from other sectors including residential sector.

5.2 Formative Years: 1970s

Among the major car manufacturing nations, the US was the first to formulate environmental policies. According to Schreurs (2003, pp. 32–33), the transformation of societal attitudes in the US towards pollution and environmental preservation began in the 1960s. Schreurs identifies this transformation as having been triggered by the rising voices on the relationship between humans and natural voices. She noted these voices as including Rachel Carson's *Silent Spring* (1964), Barry Commer's *The Closing Circle* (1971), Paul Ehrilch's *The Popular Bomb* (1968), and the influential report by the Club of Rome,[3] called *The Limits to Growth* (1972), which pointed out that the world was running out of natural resources. This new environmental awareness led to the birth of several environmental NGOs, including the Natural Resources Defence Council in 1970, the Friends of the Earth in 1969, and Greenpeace in 1971 (Schreurs 2003, p. 33). During the 1970s, these movements gradually pushed the US to the forefront in recognition as the 'environmental innovator'; its laws and institutions became models for consideration by Japan, Germany, and many other countries (Schreurs 2003, p. 254).

These green movements and the development of environmental laws and institutions were reflected in important bills regarding environmental policy on cars, such as the 1955 Clean Air Act for the reduction of smog and air pollution concerning public health. The US fuel economy regulation on cars emerged from the 1973 oil crisis. It demonstrated US reliance on cheap foreign oil and led to the creation of the Energy Policy Conservation Act of 1975. This, in turn, set the fuel economy regulations with the introduction of the CAFE standards in 1978 with 18.0 miles per gallon (mpg) target, which was improved up to 26.5 mpg by the end of the 1980s (Yacobucci and Bamberger 2007). While these progressive environmental laws on cars were passed, since they produced large vehicles with low fleets, the CAFE standard, in particular, placed negative impacts in terms of its competitiveness of the US car industry in its domestic market (Cooney and Yacobucci 2005). In contrast, it became a business opportunity for the Japanese automobile industry, which produced fuel-efficient vehicles, and gradually led to a greater share of Japanese cars in the US. This is evident in Fig. 5.1. The increasing share of Japanese cars in the US market caused trade hostilities between Japan and the US; local production by Japanese companies became a popular choice for avoiding those trade frictions. As mentioned earlier, since that time, the US has been the biggest local producer of Japanese cars for more than two decades. The local production has also encouraged the share of the Japanese automobile industry in the US market, and resulted in less than 60 % of the production of all cars and light trucks sold in the US by the 'Big Three' companies. In comparison to 1950, General Motors alone accounted for more than 50 % of the US car market. As a result, the US companies produce consumer oriented light trucks, so called 'Sports

[3]The Rome Club consisted by a group of businessman and scientists.

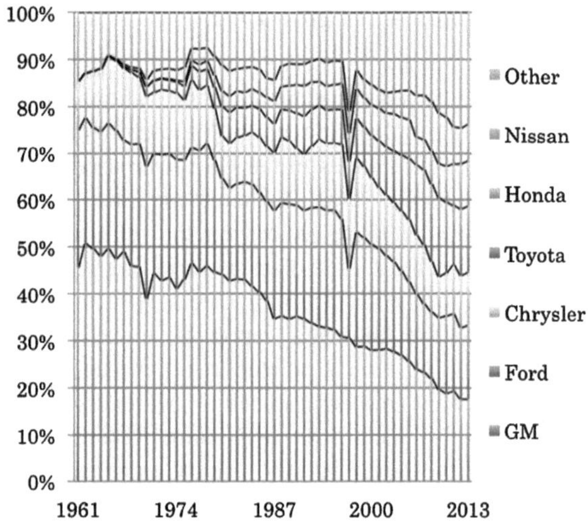

Fig. 5.1 Share of car sales in the US market by manufactures, 1961–2013. *Source* created by author based on WARDSAUTO (2014)

Utility Vehicles' (SUV), which are less fuel efficient in comparison to passenger cars (Cooney and Yacobucci 2005). Encountering the decreasing share of sales in the US market, the US automobile industry gradually opposed any increase in the CAFE regulations. The next section explains how industry lobbying led to stagnation of the US CAFE regulation.

5.3 Stagnation of the CAFE Standards: Mid-1980s

Figure 5.2 shows the stagnation of US fuel economy regulations throughout the 1980s and into the 2000s. As mentioned, the Energy Policy Conservation Act of 1975 set fuel economy regulations that led to the introduction of the CAFE standards in 1978 with an 18.0 mpg target for passenger cars. This target increased to 26.5 mpg by the end of the 1980s. Nonetheless, from 1985 to the present, the standard has remained unchanged at 27.5 mpg. For light trucks, which are sold in almost equal quantities as cars (Mikler 2010), the target was set to 17.2 mpg in 1979 and gradually increased to 24.1 mpg in 2011 (NHTSA 2004).[4] Because of the stagnant regulation, the actual fleet efficiency dropped as well. It peaked at 26.2 mpg in 1987 but has since decreased to 25.2 mpg today (Gerald and Lave 2003).

There are three main reasons that explain the stagnation of the US fuel economy regulation: industry lobby, low Congressional appropriation, and the cheap price of

[4]In fact, between 1986 and 1989, the target for passenger cars moved backwards, with 26.0 mpg as the lowest standard. This also happened for light trucks (in 1985 and in 1990).

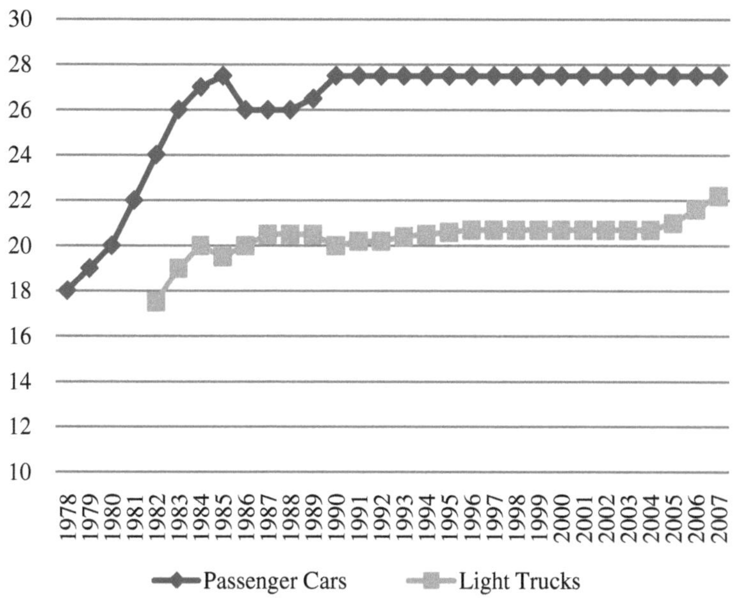

Fig. 5.2 Improvements and Stagnations of US CAFE regulations, 1978–2007 (miles per gallon). *Source* Created by author based on NHTSA (2004)

oil in the US. First, the extremely strong opposition and lobbying of the US automobile industry blocked any progress in raising the regulation. Clearly, the US automobile industry has been producing heavier vehicles compared to those of the Japanese and European industries. Furthermore, as a campaigner of Greenpeace argues, it is natural to assume that the "auto industry was making a lot of money making SUVs, and the industry saw any improvements in fuel economy would force them to improve their technology beyond 1989, and this has led them to oppose any fuel economy regulation".[5] In particular, under the CAFE regulation, compliance with the regulation is measured by calculating a sales-weighted mean of the fuel economies of a given manufacturer's product line. The penalty for non-compliance is $5.50 for every 0.1 mpg below the standard, multiplied by the number of cars in the manufacturer's new car fleet for that year. When faced with stiff civil penalties for noncompliance to the regulation, the industry supported less stringent CAFE regulations.

According to an interview conducted with a senior policy advisor, who has been working on CAFE regulation of the US Environmental Protection Agency, the main argument that the industry took to both politicians and the public was that any improvements in CAFE regulations would hurt their companies and jobs. Given

[5]Interview with Ms. Ann Mesnikoff, Director, Green transportation Campaign, Sierra Club, and Mr. Jesse Prentice-Dunn, Washington Representative, Green Transportation Campaign, Sierra Club. Interviewed at Washington DC, USA (14th June, 2012).

that the unemployment rate and stagnant economic situation in their district is a tremendous concern to them, this was a powerful argument for politicians.[6] The industries have also blocked any increase in the CAFE regulation on the basis of the climate change issue. In fact, when the George W.H. Bush administration (Republican, 1989–1993) signed UNFCCC in 1992, the US automobile industry and the US government agreed on the '1993 partnership for a new generation of vehicles', a cooperative research programme directed towards bringing high fuel efficient vehicles (up to 80 mpg) by 2003. Still, despite this agreement, the US automobile industry, together with other industries that manufacture carbon intensive products, put considerable impacts on the US climate change policy both at the national and international levels. This was especially true after the UNFCCC regime was agreed upon, in 1992, to block any progress in reducing carbon dioxide emissions. In particular, fossil fuel companies had actively lobbied at a variety of levels to prevent any measures that would involve reductions in fossil fuels (Newell and Paterson 1998, pp. 682–683). At the national level, they lobbied the US government and blocked important policy measures that were needed to strengthen the UNFCCC regime. For instance, the Clinton administration (Democrat, 1993–2001) supported the negotiation of legal commitments, at the first Conference of Parties at Berlin in 1994, to reduce greenhouse gas emissions beyond the year 2000. Nonetheless, the National Association of Manufacturers and the US Chamber of Commerce, in combination with the private electric power industry, threatened the funding for the US national climate action plan in Congress (Porter and Brown 1996, p. 61).

At the international level, the US based industries formed coalitions to block any progress in reducing carbon dioxide emissions. The most notable coalition was the Global Climate Coalition (GCC), which consisted of over 55 business trade associations and companies, including the American Petroleum Institute, British Petroleum, DuPont, Dow, Ford, General Motors, Texaco, Chevron, Mobil and Shell (Source Watch n.d.). They adopted a number of arguments to defend their interests. First, they highlighted the lack of scientific consensus and high uncertainty for taking action in emission reduction. In 1991, the fossil fuel business actors even funded their own information council, the Information Council for the Environment, with a goal to 'reposition global warming a theory, not a fact' (Levy 2005, p. 85). Second, they raised concerns about the economic consequences resulting from emission regulations in terms of unemployment. For instance, the GCC held a series of economic studies, and argued that measures to curb 20 % emissions would reduce 4 % of the US gross domestic product and 1.1 million jobs annually (Levy 2005, p. 83). Third, they engaged with the mass media to suggest the lack of evidence for climate change. Moreover, they argued that 'increased levels of carbon dioxide will increase crop production and help feed the hungry people of the world' (Source

[6] Based on interview with Mr. Jeff Alson, Senior Policy Advisor, Transportation and Climate Division, US Environmental Protection Agency. Interviewed at Ann Arbor, USA (18th June, 2012).

Watch n.d.). Fourth, they emphasised the burden for the developed countries and limited the scope for the climate change regime, given the exclusion of developing countries from emission controls. To do so, they launched an advertising campaign with the slogan 'its not global and it won't work' (Levy 2005, pp. 82–83). Because of their activity, "[t]he industry lobbying efforts of US industries were successful in securing political allies in Congress, making Senate ratification of Kyoto a very dim prospect. Federal funding for climate research has been constrained, and the US State Department opposed mandatory international GHG emission control until 1996" (Levy and Rothenberg 1999, p. 179).

Despite both national and international constraints, the Clinton administration did make progress. It accepted the need for a binding international agreement on the reduction of CO_2 emissions, and subsequently signed the 1997 Kyoto Protocol. However, it is also important to point out that the US did not ratify the Protocol and thus the Protocol had no legally binding ability for US commitment. Furthermore, at the Kyoto Protocol negotiations, the negotiation position of the US was the "most defensive among industrialized countries" due to the strong business lobbying, and therefore insisted on numerous flexibility mechanisms (Levy 2005, p. 167).

What can explain this business influence over US foreign climate policy? Scholars in the Marxist tradition argue that the business influence over US climate policies can be explained in terms of the structural power of capital, delivering from the role of the state within capitalist societies (Newell and Paterson 1998). The foundation of this argument is that the role of the states is to identify and advance the general interest of the capital, and that maintaining this capital accumulation is central to maintaining state legitimacy. Consequently, those who organize the process of capital accumulation would gain tremendous structural power over state decision-making. Regarding the politics of climate change, Newell and Paterson (1998) argue that since technologies associated with the use of oil and coal have been central to the ninetieth and twentieth century capital accumulation, fossil fuel companies were conferred significant structural power over state decision-making. Specifically, business sectors such as the coal and oil industry have been extremely dynamic in pushing the capital accumulation going, in terms of both rates and reinvestment, and innovation in production techniques. Therefore, the interests of business actors were considered a necessity for furthering capital accumulation. It follows, "state planners have therefore assumed that to keep economic growth going, planning for increased energy production and consumption was essential. Thus proposals to reduce energy consumption are still regarded as threats to economic growth" (Newell and Paterson 1998, pp. 691–695). In addition to the oil industry, the automobile industry has also been exerting its influence over US climate policy. Paterson (2007) argues that the car industry has significant implications to the growth of the US economy, as well as improving the capacity of commodity measures of mobility, and accelerating the movement of goods and people in the economy. Hence, promoting the automotive industry has played a critical role to the state in accumulating capital, and "[s]support for car thus helped to produce state power itself" (Paterson 2007, pp. 114–115).

Strong industry lobbying influenced the position of Congress in limiting the ability of the National Highway Traffic and Safety Agency (NHTSA) under the Department of Transportation, which is responsible for raising the CAFE regulatory standards. During much of this time, Congress prohibited raising the CAFE standards and passed an appropriations bill that did not allow agencies to change the standards.[7] Consequently, NHTSA decided to relax the standards for the model years 1986–1989 (Yacobucci and Bamberger 2007).

It follows that, even though they had the authority to raise the standards, the reason why the NHTSA could not raise the fuel economy standard was that appropriation by Congress to NHTSA was not enough to study raising fuel economy standards and therefore left NHTSA incapable of doing so. Congress outlawed any expenditure in each year by the Department of Transportation for any rule-making that would make any adjustment to the CAFE standards (Yacobucci and Bamberger 2007).

Environmental NGOs point out that the very reason for the lower appropriation to the CAFE regulation is because the automobile industry has powerful members in Congress and they were able to push back any potential increases.[8] The prominent example would be a bill sponsored by senators McCain and Kerry, based on the 2001 National Academy of Science[9] report, which reviewed the potential increase of the standard and advocated that car and light truck fuel economy could improve 40 % in next 10–15 years (National Academy of Science 2001). The bill would mandate a 36 mpg fuel efficiency standard by 2015, but the Senate instead voted for an amendment to charge NHTSA to study CAFE standards again (Bamberger 2002).

Another reason for the CAFE stagnation can be explained by the lack of motivation to improve the standards, due to the cheap price of oil in the US. According to an interview conducted at the Ford Motor Company, when prices were cheap, it was not just the automakers; neither the government nor the individual consumers demanded higher CAFE standards either.[10] According to Yacobucci and Bamberger (2007), p. 12, raising the price of gasoline was an unpopular discourse in the US. This preference resulted in the absence of an effective gasoline tax scheme in the country. In fact, in 1993, former President Clinton proposed the 'Btu (British thermal unit) tax', which would tax all fuel sources based on their heat content—except for renewable energy sources such as wind, solar and geothermal. The bill passed in the House, but was rejected in the Senate due to strong lobbying by the US oil industry coalition, the American

[7]Based on interview with Mr. John M. Cabaniss, Jr., Director, Environment & Energy, Global Automakers. Interviewed at Washington DC, USA (20th June, 2012).

[8]Based on interview with Ms. Ann Mesnikoff and Mr. Jesse Prentice-Dunn, Sierra Club.

[9]National Academy of Science is a nonprofit organization of scholars engaged in scientific and engineering research.

[10]Based on interview with Mr. Mark Eddie, Counsel, Environmental Affairs & Safety Regulations, Ford Motor Company. Interviewed at Washington DC, USA (21st June, 2012).

Petroleum Institute (API), and a powerful anti-climate industry coalition, the Global Climate Coalition (GCC) (Hove et al. 2002).

The US automobile industry considered that the declining price of oil would result in consumers placing less value on fuel economy and gasoline cost, but prefer bigger, more powerful and more feature-laden vehicles.[11] Therefore, rather than fuel economy, concerns about safety became the fundamental motivation of the automobile industry to produce bigger vehicles.[12] In turn, the industry has been arguing to policy-makers that if CAFE standards are to be raised, the average car size would be smaller, and this would raise the safety concern (Yacobucci and Bamberger 2007).

In addition, a backlash of government environmental regulations created an environment where any environmental progress was blocked. There was a period of intense change of environmental regulations during the 1970s, when numerous environmental laws passed, including the Clean Air Act, the Clean Water Act and the Pesticide Act. In the 1980s and 1990s, there was a new school of thought or counter balancing arguments, which argued that all these new regulations were too much and/or too quick. Therefore, beginning in the 1980s, US environmental policy became stagnant due to a backlash from industry against the complex regulatory requirements, as well as the US federal structure, which frequently blocks any important regulation enforcement (Schreurs 2003). Regarding the challenge of raising CAFE standards, many criticisms were also posed against the regulations. The claims include that the social costs of CAFE are substantially higher than the social costs of gasoline taxes in achieving oil reductions (Cardell and Dunbar 1980), CAFE forced manufacturers to produce smaller and lighter vehicles, which would result in increased traffic injuries and fatalities (Crandall and Graham 1989; Crandall et al. 1986; Shin 1990) and in doing so CAFE harmed the domestic automobile industry (Nivola and Crandall 1995).

To summarize, the automobile industry lobby blocked any increase of the CAFE regulation for more than 20 years in the US. The corporate influence over policy-makers can be explained by Neo-Gramscian theory, which claims that since big businesses, such as the auto and oil industries, are core to the capital accumulation of the US government, they have structural power over the state decision-making. This argument is evident in the low-appropriation by Congress to NHTSA to study any further possibilities of improving the CAFE regulations. Furthermore, cheap oil

[11]Based on interview with Mr. Mark Eddie, Ford Motor Company; and Ms. Julie C. Becker, Alliance of Automobile Manufactures (AAM). Interviewed at Washington DC, USA (11th June, 2012).

[12]Moreover, in contrast to Europe and Japan where gasoline prices and fuel economies of cars are high, low gasoline prices in the US may have resulted in larger size, and close to stagnant fuel economy of cars. In 2008, the gasoline price in the United States was 56 US Cents/L (34 US Cents/L in 1995), while Japanese gasoline price in the same year was 142 US Cents/L (125 US Cents/L in 1995). In Europe, the price of gasoline in 2008 was the highest, ranging from 144 US Cents/L in the United Kingdom (92 US Cents/L in 1995) to 168 US Cents/L in Netherlands (121 US Cents/L in 1995), i.e., three times more expensive than in the United States (GTZ 2009).

prices did not motivate the government or the industry to improve car fuel economy. In fact, Bill Clinton made a proposal to introduce an energy tax to increase the price of oil; however, it failed when it encountered the opposing-voices raised from the oil industry. These factors disabled any improvements in CAFE regulation until the critical changes brought in 2007. The next section describes the critical juncture, which led to significant improvements in CAFE regulation.

5.4 Critical Juncture: Massachusetts v. EPA (2007)

The critical juncture that changed the whole dynamics of US CAFE regulation was triggered by a discussion regarding whether CO_2 can be classified as an air pollutant and whether the US Environmental Protection Agency (EPA) could regulate CO_2 emissions under the authority given to it by the Clean Air Act (CCA). Such discussion in the EPA escalated during the Clinton Administration (Democrat, 1993–2001), which was inaugurated in 1992, a year after the adoption of the UNFCCC. This is evident in the memorandum from Jonathan Z. Cannon, then general council of the EPA, addressed to Carol M. Browner, then administrator of the same agency. In this memorandum, Cannon concluded that CO_2 is an air pollutant under the CCA, and "CO_2 emissions are within the scope of EPA's authority to regulate". Therefore, he urged the administrator to "date to exercise that authority under the specific criteria provide under any provision of the Act" (Cannon 1998).

In reaction to the EPA's position, on 20th October 1999, environmental NGOs and a few renewable energy companies submitted a petition to request that the EPA to regulate Greenhouse gas (GHG) emissions ($CO_{2,}$ methane, nitrous oxide and hydro fluorocarbons) from motor vehicles under its authority conferred by the Clean Air Act (CCA) (International Center for Technology Assessment 1999).[13] Nevertheless, the EPA did not make any reply to the petition until 2003. This can be partially explained by the timing of its submission. Given that President Clinton's time in office was only until January 2001, there was not enough time left for the EPA to take any action. Furthermore, even though the Clinton administration made progress in terms of international climate negotiations, when it came to domestic climate policies, exertions to make improvements encountered tremendous opposition from industry.

[13]Petitioners were consisted by 14 environmental NGOs and 4 industry associations which promote renewable energies, and one bi-partisan group of members of Congress, called Environmental and Energy Study Institute. Environmental NGOs include Center for Technology Assessment, Alliance for Sustainable Communities, Bio Fuels America, Earth Day Network, Environmental Advocates, Friends of the Earth, Full Circle Energy Project, The Green Party of Rhode Island, Greenpeace USA, National Environmental Trust, Network for Environmental and Economic Responsibility, New Jersey Environmental Watch, New Mexico Solar Energy Association, Public Citizen,.

Accordingly, under the new Republican administration, led by President George W. Bush beginning in January 2001, a newly appointed general council, Robert E. Fabricant, concluded in his memorandum addressed to the administrator of the EPA on 28 August 2008 that the "CAA does not authorize EPA to regulate for global climate change purposes. Accordingly, CO_2, and other GHG cannot be considered air pollutants subject to the CAA's regulatory provisions for any contribution anthropogenic GHG emissions may make to global climate change" (Fabricant 2003). On the same day, the EPA denied the petition, reasoning that CAA does not grant the agency the authority to regulate CO_2 emissions. It states:

> EPA concludes that it cannot and should not regulate GHG emissions from U.S. motor vehicles, under the CAA. Based on a thorough review of the CAA., its legislative history, other congressional action and Supreme Court precedent, EPA believes that the CAA does not authorize regulation to address global climate change. Moreover, even if CO_2 were an air pollutant generally subject to regulation under the CAA, Congress has not authorizes the Agency to regulate CO_2 emissions from motor vehicles to the extent such standards would effectively regulate car and light truck fuel economy, which is governed by a comprehensive statue administered by DOT (EPA 2003).

This denial caused the critical juncture of the US CAFE regulation, led by each state such as California. Followed by the EPA's denial of the petition, 12 states including Massachusetts and California, several cities such as the city of New York, as well as environmental NGOs prompted a lawsuit, *Massachusetts v. EPA 549 U. S.* 497, in the DC Circuit and later in the Supreme Court during 2006 and 2007.[14] They argued that since these emissions can 'reasonably be anticipated to endanger public health or welfare', the Clean Air Act requires that the EPA regulate CO_2 and other GHG emissions (US Supreme Court 2006).

In turn, respondents claimed that the CAA does not authorize the agency to regulate GHG emissions, and even if it did, the agency does not prefer to use that authority until the causes, extent and significance of climate change and the potential options become more certain (US Supreme Court 2006). Specifically, they argued that since the Act was written in the 1970s, Congress did not anticipate using the CAA to regulate GHGs during the time of the passage of the bill. These respondents include the Environmental Protection Agency, the Alliance of Automobile Manufacturers and other industry associations, as well as automobile manufacturing states such as Michigan and oil producing states such as Texas.[15]

[14]Other petitioners are Connecticut, Illinois, Maine, New Jersey, New Mexico, Oregon, Rhode Island, Vermont and Washington, the cities of Baltimore and Washington D.C., the territory of American Samoa, and research institutes Center for Biological Diversity, Center for Food Safety, Conservation Law Foundation, Environmental NGOs, Environmental Defence, Friends of the Earth, Greenpeace, International Center for Technology Assessment, National Environmental Trust, Natural Resources Defence Council, Sierra Club, Union of Concerned Scientists, and the U. S. Public Interest Research Group (Cornell University Law School 2007).

[15]Other respondents include Alaska, Idaho, Kansas, Nebraska, North Dakota, Ohio, South Dakota, Texas, and Utah, and industry group National Automobile Dealers Association, Engine Manufacturers Association, Truck Manufacturers Association, CO_2 Litigation Group, and Utility Air Regulatory Group (Cornell University Law School 2007).

During the Court discussions, the Court held that the CAA does give the EPA the authority to regulate tailpipe emissions of greenhouse gasses, as CAA provides that "[t]he Administrator shall by regulation prescribe in accordance with the provision of this section, standards applicable to the emission of any air pollutant from any class or classes of new motor vehicles or new motor engines, which in his judgement cause, or contribute to, air pollution which may reasonably be anticipated to endanger public health or welfare" (EPA n.d.). Furthermore, the majority opinion of the Court commented that CO_2 fit well within the CAA's definition of air pollutant, which, in definition, "any air pollution agent or combination of such agents…substance or matter which is emitted into or otherwise enters the ambient air" (EPA n.d.).

Conclusions were achieved on 2 April 2007. The Court rejected the EPA's argument and held that section 202 of the CAA authorized the EPA to regulate emissions from new motor vehicles based on their possible climate change impacts. Furthermore, section 202 does not authorize the EPA to inject policy considerations into its decisions whether to so regulate (US Supreme Court 2007).

This Supreme Court decision created a big potential change in US fuel economy regulation and its climate policy in general in the following three ways (US Supreme Court 2007). First, since this judgement is the first pronouncement on climate change in the US, this means that "the finding of standing likely will be pivotal to the fortunes of plaintiffs in other climate litigation" (Meltz 2007). Second, it suggests that EPA, along with NHTSA, also has an authority to improve any CAFE improvements under the CAA. This means that US CAFE regulation will be motivated to raise its standards not only in terms of energy conservation, but also in concerns of climate change issues. Any future US climate policy can directly influence CAFE standards. Third, and most important, based on the decision, on 30th June 2009, the EPA granted California the authority to implement its own GHG emission reduction standards for new passenger cars (see the section below for more details). In fact, California had been arguing that the CAA, rather than the federal government, gives California special authority to enact its own air pollution standards for passenger vehicles. To do so, California passed its own legislation in 2002 called 'Assembly bill 1493' (Pavely), and requested that the EPA implement its own GHG standards for passenger vehicles in December 2005 in conjunction with CAA (as known as a 'waiver request'). California's adoption of its own standards for passenger vehicles caused the 'California effect', in which the federal standards converged with those of California. The next section describes how, in 2007, this critical juncture gradually changed the stagnant CAFE regulation for the first time in the previous 20 years.

5.5 The 2007 Energy Security and Independence Act

Under the presidency of George W. Bush (Republican, 2001–2009), the US withdrew from the Kyoto Protocol in 2001. Bang et al. (2005) identify three reasons for this withdrawal. Concern over negative economic consequences among decision-makers was first. Second, with the exclusion of developing countries from emission controls, the burden for the US was perceived to be heavy. Third, with the anticipated negative economic effects, the "governing system of checks and balance between the legislative and the executive branches resulted in a continuous political majority against ratification of the Protocol" (Bang et al. 2005, p. 286). Since then, the US climate policy became deadlocked—despite several attempts at proposing climate bills in Congress, including the joint bill proposed by Senator John McCain (Republican) and Senator Joseph Lieberman (Democrat), which proposed a mandatory cap and trade system for GHGs in 2003, 2005 and 2007. All of these bills failed to gain enough votes in the Senate.

In contrast to adherent US climate policies, which did not make any improvements during that period; the mandatory fuel economy standard, which required the industry to improve fuel economy by 40 % (35 mpg) by 2020, was legalized under the Energy Independence and Security Act in 2007 (White House 2007). Indeed, a number of important initiatives regarding the automobile industry were introduced during the presidency of George W. Bush. The primal rationale of these bills was concern with the enhancement of the competitiveness of the automobile industry, reducing dependence on foreign oil and thus strengthening US energy security. The first of these initiatives was the '2002 Freedom CAR Partnership',[16] which aimed to develop technologies such as fuel cell vehicles and "will establish the United States as a global leader in environmental and energy technologies and will be a key to ensuring future US competitiveness" (US Department of Energy 2002). In the following year, this initiative was further strengthened by the Hydrogen Fuel Initiative in which the Bush administration sought $1.2 billion to improve hydrogen-powered fuel cells and the hydrogen infrastructure. In addition, the 2005 Energy Policy Act increased the share of bio-fuels as well as the share of Flex Fuel Vehicles, that is Internal Combustion Engine vehicles running on gasoline blended with either ethanol or methanol fuel. Furthermore, to satisfy the stable and large scale provisions of alternative fuels, the Bush administration launched the 'twenty-in-ten initiative', which aimed to improve vehicle fuel economy and increase alternative fuels, while reducing US gasoline usage by 20 % by 2017. What seems apparent from these initiatives is that the Bush administration aimed to enhance the industry competitiveness and address energy security issues by investing in longer-term technologies such as hydrogen and flex fuel vehicles, without imposing mandatory fuel economy regulation. This strategy is consistent with the industry preference. Prior to dismissal of the GCC, it stated "[t]he Bush administration will soon announce a climate policy that is expected to rely on the development of new technologies to reduce greenhouse gas emissions, a concept strongly supported by the GCC" (Source Watch n.d.).

The question then arises, why did the Bush administration need to introduce improvements in CAFE regulation, which had been stagnant for almost for 25 years? There are three reasons for this historic change. The first and most

[16]The Freedom Car Partnership was announced in 9th January 2002 by US Secretary of Energy, DaimlerChrysler, Ford and General Motors.

important reason could be identified as the Supreme Court decision and California's role, as mentioned in the section above. In contrast to the federal level, state level environmental regulation of the automobile industry has evolved significantly. The best example of stringent state-level regulation of cars is California's LEV regulation in the 1990s, which required all car manufacturers to include zero-emission vehicles as a small percentage of their total sales in the state. Although the industry's lobbying activities gradually weakened the regulation, it mandated the industry to include 2 % ZEVs in 1998 and 10 % in 2003, including advanced-technology partially zero-emission vehicles such as hybrid cars. Furthermore, California successfully established Assembly Bill No. 1493 (so-called 'Pavley Law') in 2002, which entered into force on 1 January 2006. It set fuel economy standards for two separate car categories from 2009 to 2016. It requires car manufacturers to achieve 323 g/mile in 2009, 205 g/mile in 2012, and 172 g/mile in 2016 for passenger cars. This standard was much stricter than the federal level target set by the CAFE standard.

Since it placed the 'California effect' on a number of US domestic environmental regulations during the 1990s, California's emission standards changed the whole dynamics in the exact same way. The 'California effect' first appeared when California set its own stringent emissions standards regulating Nitrogen Oxide (NOx) and Sulphur Oxide (SOx) from cars under the amendment of the 1970 Clean Air Act. As a result, Congress brought national emission standards up to California's level and permitted California to raise its own emission standards even more. Twelve eastern states followed California's standard in 1994, and federal level emission regulations once again caught up with California's standard (Vogel 1997). It follows that California's own fuel economy regulation implied that the California regulatory standard would spill over to other like-minded states and hence create an environment where the federal level regulation is likely to catch up with that of California.

Moreover, it created a situation where there were overlapping authorities over the fuel economy regulations. In addition to the NHTSA, which was conferred authority over the fuel economy regulation by the Energy Policy Conservation Act of 1975, the EPA could also regulate car CO_2 emissions under the Clean Air Act. In fact, the EPA approved California's own CO_2 emissions standards for cars in 2009. This further enabled California to step forward, and its like-minded states to follow California's standard.

Since California is such a large market for the automobile industry to ignore, the situation of multiple regulatory obligations became a tremendous concern to the industry. Accordingly, in December 2004, the American Automobile Manufacturers Association (AAM) filed a suit in the district court against the California Air Resources Board for inappropriate authority. Against the expectations of the industry, the court approved California's authority over the fuel economy, which implies that other states can choose to set their own standards, just like what happened in the 1990s, as a complement to the federal CAFE standard (Inoue 2008). The recent US multiple potential fuel economy regulation is shown in Fig. 5.3.

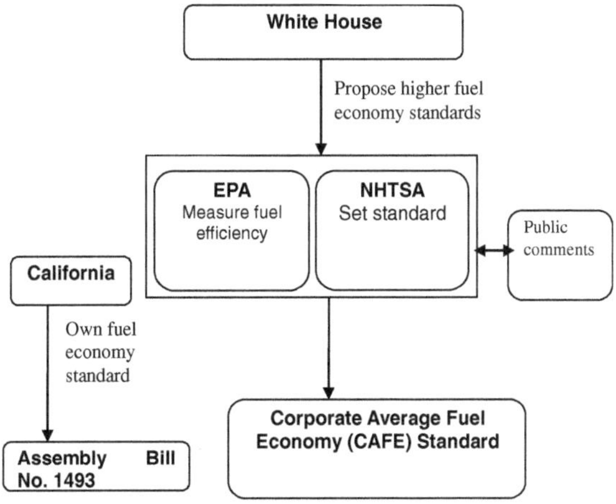

Fig. 5.3 US multiple potential fuel economy regulation. *Source* Iguchi and Hillman (2012)

By 2007, 40–50 % of the new car market was covered by states that had adopted California's standard. The price of gasoline started to rise as well. This inevitably applied pressure to the automobile industry, since they have to clear these standards to sell vehicles in the states that adopted California's standard. In turn, the industry supported the national standards, which resulted in Bush's 35 mpg by 2020 target. According to the interview conducted to Ford:

> California adopted their AB 1493 GHG standards in 2004 as a result of concerns about climate change, and these standards were essentially fuel economy standards by another name. The AB 1493 standards were a tremendous concern for the industry, because they would have required manufacturers to manage the GHG emissions of their vehicle fleets on a state-by-state basis. State-by-state regulation would have added great complexity to manufacturers' plans for product development and distribution. It also meant that manufacturers would likely have been forced to restrict sales of some products in different states because of GHG regulations. This would reduce vehicle sales, harm dealerships in the affected states, and annoy consumers. The auto industry believed (and continues to believe) that the California GHG regulations were prohibited by the federal CAFE law. Therefore, it was a legitimate business choice for the auto industry to try to overturn the California regulations through litigation. No one wanted a 'multi-state scenario.[17]

According to an interview conducted with a campaigner supporting raising the fuel economy standard since 1989, the role of the NGOs was instrumental in helping California set its own fuel economy standards. The reason why the NGOs pushed California to adopt its own emission standards was that the automobile industry had been blocking any fuel economy increase in Congress. Therefore, as an alternative, they launched a campaign in California to pass the Pevely law, which was the first emission standard. Then, knowing that the mid-states are automobile manufacturing states, the environmental NGOs campaigned in both the West coast states and the East coast states to adopt California's rule. Consequently, since

[17]Quoted from interview with Mr. Mark Eddie, Ford Motor Company.

several states adopted California's standard, this led to changes in the political dynamics as well as the strategy of the industry.[18]

The second reason for the increase in the fuel economy standard was the rise of the gasoline price, starting in 2005, which motivated increasing the CAFE standards. Although the gasoline price was around $1.7, beginning in 2005, it rose and reached $3 a gallon, and then reached $4 a gallon in 2008 (US Energy Information Administration 2013). This implies that US consumers were now paying gasoline prices similar to earlier prices paid by European and Japanese consumers. Whereas, for 25 years, nobody had cared much about the cost of gasoline, the years 2005 and 2006 became a time of major change. Many consumers and politicians started to express concern over how much gasoline cost.

As the name suggests, the Energy Independence and Security Act aimed to reduce dependence on Middle Eastern oil imports. Such discourse became popular among a group of stakeholders called 'neoconservatives', comprised of military generals and ex-generals, within the supporters of the Republican Party. The most notable groups of neoconservatives included a coalition called 'Set America Free', comprised of members including former Republican candidate Gary Bauer, powerful neoconservative journalists such as Frank Gaffney and Daniel Pipes, the former Central Intelligence Agency director Jim Woolsey, as well as two environmental activists Deron Lovaas and Bracken Hendricks (Set America Free n.d.). Their major claim was to end oil dependency for energy security purposes. They argued that since a large amount of gasoline is consumed through the use of cars, they strongly urged improving fuel economy rates for automobiles, as well as increasing the rate of biofuels and hybrid electric cars sold in America (Set America Free 2004). Triggered by the energy security concerns, this series of discussions eventually led Congress to adopt requirements to change the CAFE standards for the 2012–2016 model years.

Encountering high gasoline prices, political support increased within the Republican Party and the voice of anti-CAFE regulation gradually became smaller and smaller. A good example would be Congressman John Dingel, from Michigan, who has been strongly opposed to increasing any CAFE raise. Arguing that it would harm the domestic automobile industry, he proposed a carbon tax that would increase the gasoline tax by 50 cents, instead of raising the CAFE regulation. It failed to pass in Congress.

The third reason lies in the decline of competitiveness of the US automobile industry and its influence over Congress. The rise in gasoline prices meant a shift in consumer preference to more fuel-efficient vehicles. As previously mentioned, the American automobile industry concentrates its production on larger vehicles such as SUVs, which are low in fuel efficiency. Their competitive index has been the *safety* of the vehicle, not the fuel efficiency. Therefore, during the economic downturn and the increase in gasoline prices, the American automobile industry

[18]Based on interview with Mr. Dan Becker, Director, Safe Climate Campaign, Center for Auto Safety. Interviewed at Washington DC, USA (20th June, 2012).

was vulnerable, and allowed the Japanese automobile industry to expand their share in the US market (Freedman and Blair 2010).

While these manufacturers produce and sell most of the vehicles within the domestic market, its share in the market has been declining since the late 1970s. As hitherto noted, US car manufacturers dominated more than 70 % of vehicle sales in the US market until the mid-1990s, and then gradually dropped its share to the present level. In 2010, the sales share of the total US car manufacturers in the US market represented less than 50 %.

In contrast, Japanese car manufacturers, especially Toyota, gradually increased their share since the late 1970s, and in 2010 represented 30 % of the total US car sales. This suggests that the introduction of the US CAFE regulation in the early 1970s actually triggered the space for Japanese car manufacturers to produce fuel-efficient cars in the US market. Furthermore, regardless of the 2007 Energy Security Act, which dramatically enhanced the US CAFE regulation, the share of the US auto manufacturers declined in comparison to the increase of the Japanese auto manufacturers. Table 5.1 shows the motor vehicle production and sales by major US car industry since 2007. It shows that although the share of the US automobile industry declined in the US market, its profit came from sales in the domestic market. This clearly demonstrates the weak competitiveness of the US automobile industry.

Experiencing the rise of the price of oil and the changes in consumer demands to purchase more fuel efficient vehicles, the US based automobile industry's power has been significantly weakened. They have argued for 25 years that if the CAFE standard is to be raised, their plants will be shut down, it will hurt their sales, and they will have to lay off autoworkers. Nonetheless, even though the standard did not change, they still had to close plants and lay off workers. A campaigner of the environmental NGO observes that the automobile industry lost its credibility in Congress because they had to shut down their manufacturing plants in several states, not because of the rise of CAFE standards, but because they were not competitive enough to produce low-fuel efficient vehicles.[19]

If we look at the money spent in lobbying for the automobile industry, the US automobile industry spent a considerable amount on lobbying from 2006 to 2007, the year in which former President Bush announced the introduction of a new CAFE standard. The US automobile industry spent approximately US$0.7 billion on lobbying activities between 1998 and 2012; this is the 18th largest spender among US industries. This amount increased since 1992, the year that UNFCCC was adopted, and it continued to increase until 2007. There were gradual decreases in the expenditure from 2007 to 2010, probably due to the financial crisis. Still, the expenditure once again increased in 2011, the year that President Obama announced the stringent fuel economy regulation. Table 5.2 shows the annual lobbying expenditures by the US automobile industry from 1998 to 2011. It is notable that despite the fact that the industry spent a considerable amount on lobbying, a mandatory rise of the CAFE regulation to 35.5 mpg (172 g CO_2/km) by

[19]Based on interview with Mr. Dan Becker, Center for Auto Safety.

Table 5.1 Motor vehicle production and sales by major US car industry by region, 2007–2010

Year/company		GM	Ford	Chrysler
2007	Production	**US: 2,801,450 (30 %)** EU: 1,928,323 (21 %)	US: 2,174,599 (35 %) Asia: 369,113 (6 %) **EU: 2,303,944 (37 %)**	**US: 1,651,285 (65 %)** EU: 68,913 (3 %)
	Sales (billion US dollar)	**North America: $112.4 (62 %)** Europe: $37.3 (20.6 %) Asia-Pacific: $21 (11.5 %)	**North America: $70.4 (45.5 %)** Europe: $36.3 (23.5 %) Asia-Pacific: $7 (4.5 %)	–
2008	Production	**US: 2,285,733 (28 %)** EU: 1,643,717 (20 %)	US: 1,602,011 (30 %) Asia: 306,379 (6 %) **EU: 2,142,498 (40 %)**	**US: 1,106,028 (58 %)** EU: 28,207 (1 %)
	Sales	**North America: $86.1 (57.8 %)** Europe: $34.6 (23.2 %) Others: $24 (16.1 %)	**North America: $53.3 (41.7 %)** Europe: $37.6 (29.4 %) Asia-Pacific: $6.5 (5 %)	–
2009	Production	**US: 1,185,661 (18 %)** EU: 1,137,853 (18 %)	US: 1,390,870 (30 %) Asia: 478,520 (10 %) **EU: 1,660,017 (35 %)**	**US: 481,183 (20 %)** EU: 5,376 (0.5 %)
	Sales	**North America: $56,617 (54.1 %)** Europe: $24,031 (22.9 %) Other: $14.7 (14.1 %)	**North America: $49.7 (47.8 %)** Europe: $28.3 (27.2 %) Asia-Pacific: $5.6 (5.3 %)	–
2010	Production	**US: 1,719,541 (20 %)** EU: 1,246,527 (15 %)	**US: 1,690,973 (34 %)** Asia: 501,668 (10 %) EU: 1,304,296 (26 %)	**US: 838,497 (53 %)** EU: 5,497 (6 %)
	Sales	**North America: $83 (61.2 %)** Europe: $24 (17.7 %) Other: $21.4 (15.8 %)	**North America: $64.4 (57.9 %)** Europe: $29.4 (26.4 %) Asia-Pacific: $ 7.3 (6.5 %)	–

Bold indicates the largest share. *Source* OICA correspondence survey (2007, 2008, 2009, 2010; General Motors 2007, 2010; Ford Motor Company 2007, 2008, 2009, 2010), modified by the author

2020 was passed in Congress by *unanimous* vote. Considering that the automobile industry had more than 41 senators successfully vote to block any increases in fuel economy standards, the unanimous votes by the senators also indicates the declining influence of the US automobile industry.

Table 5.2 Annual Lobbying by Automobile industry, 1998–2011 (million US Dollars)

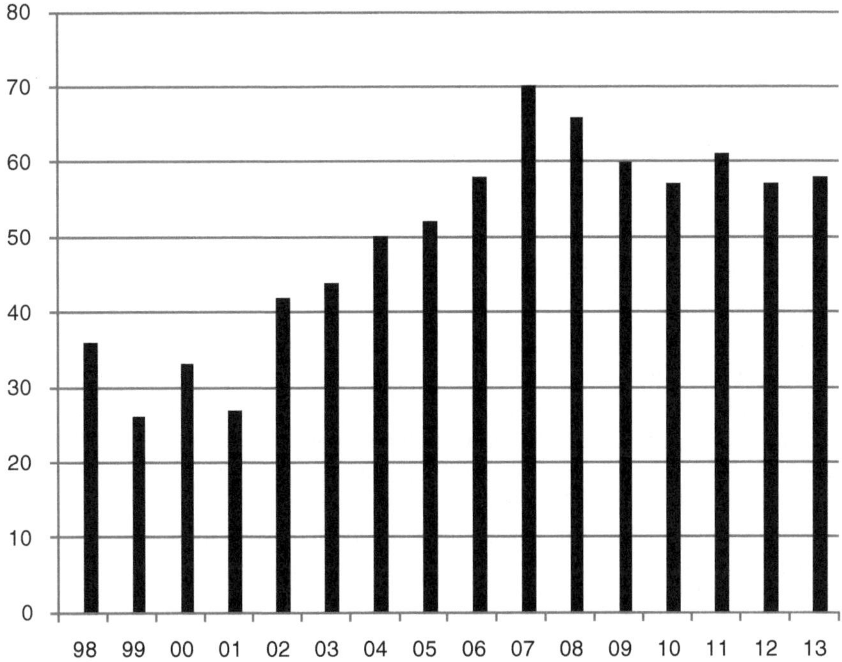

Source OpenSecrets.org (n.d.), based on Senate Office of Public Records

To summarize, all dynamics in the context of the US fuel economy regulation changed with the Supreme Court decision to allow regulating CO_2 emissions in the context of the Clean Air Act. This led the state of California to set its own standard, which was far more stringent than the federal standard. Therefore, the industry primarily supported for a unified standard between federal and state levels because of an emergence of multiple sources of the regulation emerged. Furthermore, the increasing gasoline price during the 2000s led to two things. First, it decreased the number of sales of larger vehicles and therefore the automobile industry eventually had to shut down its plants. Second, the neoconservatives argued for the reduction of dependence on foreign oil and in doing so created another stream of support for the CAFE increase. The decline of the US based automobile industry triggered a decline of their credibility in Congress, which is a crucial source of votes for Democrats. In contrast to the decline of political power over Congress by US based manufacturers, Japanese manufacturers increased their political influence and that created another source of force for the advancement of the CAFE standard. As a result, the Energy Security and Independence Act mandated the further increase of the CAFE regulation. The next section looks at the political dynamics that caused the convergence of the fuel economy regulatory standard in the US up to the level of that of Europe and Japan.

Table 5.3 The proposed fleet average CO_2 emissions (g/km) of US passenger cars

Year	Federal	California
2011	228	221
2012	205	205
2013	198	198
2014	191	191
2015	181	181
2016	172	172

Source Based on ICCT (2011), modified by author

5.6 Target for 2015 and Beyond

This section looks at the political dynamics that led to the regulatory convergence of the fuel economy for passenger cars among the US, Europe and Japan under the Obama Administration. President Obama came into office in January 2009. This was an important year for international climate policy as parties to the UNFCCC were expected to reach an agreement on the post-2012 framework at the 15th Conference of the Parties (COP15) held in Denmark in December 2009. Prior to the COP15 meeting, President Obama announced the intent to reduce the 17 % reduction of US CO_2 against the 2005 level by 2020, a 43 % reduction by 2030, and an 83 % reduction by 2050.[20]

Furthermore, President Obama decided to advance the CAFE regulation set by the Bush administration by proposing to shorten the target of 35.5 mpg (172 g CO_2/km) to be achieved by 2016 (White House 2009a, b). In the presidential speech in May 2009, President Obama stated:

> In the past, an agreement such as this would have been considered impossible… That is why this announcement is so important, for it represents not only a change in policy in Washington, but the harbinger of a change in the way business is done in Washington. As a result of this agreement, we will save 1.8 billion barrels of oil over the lifetime of the vehicles sold in the next five years. And at a time of historic crisis in our auto industry, this rule provides the clear certainty that will allow these companies to plan for a future in which they are building the cars of the 21st century (White House 2009a).

As a result, what occurred was a synchronization of the fuel economy standards on both the federal and the California level from 2012 until 2016 (ICCT 2011). Table 5.3 shows the synchronization of federal and California standards. Here, we can observe that the 'California effect' explicitly influenced the national fuel economy regulation.

In July 2011, President Obama made a proposal to improve CAFE regulation up to 62 mpg for cars and 44 mpg for trucks by 2017–2025 (CNN 2011). In doing so, he appointed the EPA and the NHTSA to conduct a study whether the proposed

[20]After the COP15 meeting, domestic bill that mandates these emission cuts were failed due to the strong oppositions by the Congress.

target is achievable by the US automobile industry. Here, again, California's role was important, since it had already started its own standard for 2017 and after, and the automobile industry attempted to stop California from implementing its own standard, and thus supported a uniform standard by the federal government. As a result, both the EPA and NHTSA coordinated their efforts to conduct a study to evaluate new fuel economy regulations up to 2025 (EPA 2011). Therefore, the EPA and the NHTSA proposed stringent targets for the CAFE standard from 2017 to 2025, which call for 47–62 mpg (corresponds to 99–131 g CO_2/km) by 2025 at the most stringent level (EPA 2010).[21]

The conclusion was reached on 15 October 2012, when the EPA and NHTSA announced the new fuel economy standard of 54.5 mpg between 2017 and 2025. This target was preceded by a multi-party agreement, promoted by the White House, the automobile industries including both American and foreign nations, the United Auto Workers, California and other interested parties (EPA and NHTSA 2011).

In deciding on the 54.5 mpg target, there was a range of negotiations to decrease 3–6 % GHGs annually, ranging from 45 to 62 mpg calculated and proposed by the EPA and the NHTSA. The negotiation came down to roughly a 5 % annual decrease in emissions, which is about 55 mpg, but they also take into account the target level of pick-up trucks, which weakens the target a small amount, and 54.5 mpg was agreed upon as a final target.[22] In addition, unlike the EU's fuel economy target for 2020, which has a constituency with the EU's overall CO_2 reduction target, the US fuel economy target for 2025 does not have a direct linkage with any national or international climate targets.[23]

Following six factors helped to push the US CAFE regulation up to the level of the European and Japanese standards. They are: concerns to enhance competitive of the US automobile industry; Obama's leadership, the positive role played by the EPA; changes of position by the Union of Automobile Workers (UAW); automobile industry's strategy to avoid multiple sources of regulations; and US automobile industry's decline of political influence over the Congress.

Firstly, the most important factor, in addition to these enabling environments and political changes, involves competitiveness issues for the US automobile industry. According to an interview conducted with a senior policy officer at the US EPA, the target was primarily decided upon to enhance market competitiveness. Furthermore, "there were realizations both high levels of government on the one hand, and high level within automobile industry on the other hand that the industry is much more global in the future and much less regional specific markets where US market sells 25 miles per gallon (mpg) vehicles while Europe and Japan sells 40 mpg vehicles. In other words, there is recognition that the differences between markets are smaller

[21]They set four different scenarios for the fuel economy standard until 2025, entailing three to six per cent improvement per year.

[22]Based on interview with Ms. Mesnikoff and Mr. Jesse Prentice-Dunn, Sierra Club.

[23]Based on interview with Mr. Jeff Alson, US EPA.

and smaller".[24] In fact, the US automakers, realize that if they want to be successful, they cannot rely exclusively on the US market. They have to compete with rivals over the emerging markets of Brazil, Russia, India, China and South Africa (BRICS), and they cannot do that with 20 mpg. According to an interview conducted with the vice president of the Environment Department of AAM, the competitiveness issue was central in formulating the 2017–2025 CAFE regulation, and "[o]ne of the reasons that the industry has agreed on this is influence of European and Japanese standards, and we need to stay competitive globally. Therefore, the members felt that there was a need for global platforms".[25]

In order to promote the enhancement of the competitiveness of the US automobile manufacturers, because of the increasing concerns of the Big Three companies, the new CAFE regulation aimed for 2016–2025 introduced several changes to the structure of the standards. Namely, under the old structure in which one single standard applies to all companies whether it produces smaller or larger vehicles, the same standard applies to all of the companies, regardless of their products. One of the new measures of the new CAFE regulation is the so-called 'footprint' based regulation. Unlike the old 'one size fits all' structure of CAFE regulation, the footprint standard sets the standard according to the vehicle size (EPA 2012).

The footprint-based standard would improve the actual fuel economy of cars and thus enhance its competitiveness in the following ways: first, it avoids the manufacturers achieving the standard by simply using lighter materials, and not making endeavours to improve their fuel economy. If one would put a lighter material into the car but keep the same size of the car, the standard stays the same, and therefore one would benefit simply by using lighter materials; second, it will safeguard against industry claims that bigger cars are safer than smaller cars, and therefore an increase in CAFE regulation would threaten the safety of the American public. In comparison to a weight-based standard, which would encourage manufacturers to produce smaller cars, a footprint based standard would not motivate the manufacturers to produce smaller cars since the regulation is differentiated according to the size of the vehicle.

The second important factor was having a new president in office and the difference in his leadership. A senior advisor in the US EPA expresses, "if different President came in, this was probably not happened" (see footnote 24). Indeed, when President Obama was a senator, he was extremely interested in oil and carbon issues. For instance, in July 2006, he sponsored a bill, 'S.3694 Fuel Economy Reform Act' to raise CAFE by 4 % per year until 2018, which did not come to a vote but was read twice and referred to the Committee on Finance (Library of Congress 2006). Right after he was elected, he announced that 55.5 mpg (103 g/km) by 2020 was too slow, and assigned the EPA and NHTSA to work together to propose a more aggressive fuel economy to achieve the new standard by 2016. In July 2011, the president again announced new fuel economy standards of 47–62 mpg by 2025. Consequently, the

[24]Interview with Mr. Jeff Alson, US EPA.

[25]Interview with Ms. Julie Becker, AAM.

EPA and the NHTSA proposed 54.5 mpg as the target in 2011, which is double the number when President Obama was elected.

In this process, the positive role played by the EPA could be identified as the second reason. As previously mentioned, the Bush Administration's EPA did not make any progress, although the Court decision on *Massachusetts v. EPA* clearly states that the EPA has the authority to regulate GHG emissions since CO_2 emissions are defined as air pollutants under the Clean Air Act (McCarthy 2009). Instead, it argued that 'the Clean Air Act is a deeply flawed and unsuitable vehicle for reducing greenhouse gas emissions' (EPA 2008, p. 44356). Therefore, as soon as the Obama Administration took command of the White House, it urged the EPA to review the current state of car CO_2 emissions. It did so by appointing the EPA to work jointly with NHTSA—as compared to just the NHTSA, which only has a couple hundred staff members. The EPA brought a fresh perspective, with better manpower, to the study on new CAFE standards for new cars and light trucks.

Consequently, in April 2009, the EPA proposed that GHGs do indeed endanger both public health and welfare and concluded that car CO_2 emissions largely contribute to such endangerment (EPA 2009a). Based on the calculation of the cost associated with the fuel economy improvement, they concluded that a 55.5 mpg target by 2025 could be achieved cost-effectively. This was a big step in the process because even though the industry attempted to weaken the number reasoning for their technological capabilities, with the comprehensive and detailed data, EPA could defend itself from these types of arguments. Furthermore, this meant that for the first time, the EPA made a step forward in defining how the Clean Air Act could be used (McCarthy 2009). It also implied that with the authority conferred by the Air Pollution Act, the EPA could take several steps to regulate GHG emissions from not only the road transport sector, but other transport sectors as well (McCarthy and Yacobucci 2013).

This EPA's proposal was later discussed and agreed upon between the US government (under the leadership of President Obama), the EPA (which has the authority to develop its own GHG emission standards for cars under the Air Pollution Act), the NHTSA (which was originally conferred the authority over CAFE regulation under the Energy Conservation Act), California (which set its own standard) and fellow states, the automobile industry, United Auto Workers (which represents workers in the automobile industry) and environmental groups (EPA 2009b).

Third, with the change of the completely new dynamics—the bailout of the US automobile industry,[27] President Obama's new leadership, new EPA's positive roles and scientific evidence that proved that the new proposed target for 2025 is

[27]According to interview conducted with JAMA-US, they observed that the influence of the U.S. based auto industry declined, especially after the bailout - Since the federal government helped GM and Chrysler with a significant amount of taxpayer's money to stay them alive. This eventually led the US-based auto industry to say no to what the President wanted. Based on interview with Mr. Masami Tanaka, Deputy General Director, Japan Automobile Manufacturers Association Washington Office, and Mr. Ronald Bookbinder, then-Director, Government Affairs, Japan Automobile Manufacturers Association Washington Office. Interview conducted at Washington DC, USA (20th June, 2012).

achievable—there has been a change in the position of the UAW, the workers' union in the automotive sector, to support for stringent CAFE regulation. Since it is one of the major sources of votes in auto-manufacturing states such as Michigan and Ohio, the UAW has been an important political player; therefore, the UAW position has largely influenced politicians on the CAFE standard (Boyle 1998).

Although they originally supported CAFE standards in the 1970s, during the period of low gasoline prices, they joined the industry to oppose the higher standards. This was because they had the same concerns as GM, Ford and Chrysler—that raising standards were harmful to US based vehicles, which were due to their concerns that higher fuel economy regulation would result in the loss of jobs (Nordhaus and Shellenberger 2005).

However, faced with the decreasing share of the US based automobile industry and losing employment, the UAW changed its position and recognized that US companies needed to produce more fuel efficient cars that can compete with Japanese and European auto industries. Furthermore, to so fairly, they recognized that one uniform standard was needed for everyone in order to compete with the Japanese and European manufacturers (UAW 2012).

This change of UAW position is evident in 2012 public hearing regarding to 201–2025 fuel economy regulation hold by the EPA and NHTSA, Bob King, the president of UAW stated as follows:

> Its an honor to be here this morning on behalf of our membership to voice UAW's full and strong support for the proposed rules, regulating greenhouse gas emissions and fuel economy. The proposed rules are sensible, achievable and needed. They are good for the automobile industry and its workers, good for the broader economy, good for the environment and good for our national security…One important reason we are so confident that the industry's future – in the industry's future is that we are excited about the new green technologies that are being developed in the United States and produced in UAW-represented facilities…A second, more fundamental reason is because the technology needed to improve efficiency and reduce pollution represents additional content on each vehicle. That additional content must be engineered and produced by additional employees (King 2012).

It follows that the bailout of the automobile manufacturers proved that they had to shut down manufacturing plants not because of higher standards of fuel economy regulation, but because of the lack of competitiveness that primarily resulted from the lower fuel efficiency vehicles they produced.

The fourth reason is that the industry strategy to avoid the divergent regulations continued to play a critical role in pushing for the 2025 target. As mentioned, to avoid the potential risk of losing in court and facing state-by-state enforcement, the industry supported one national programme standard of fuel economy regulation. After the introduction of the 2012–2016 regulation, the industry was still concerned about the possibility that California and other states would revert to state-by-state standards for 2017 and beyond.[28] Since the automobile industry avoided multiple fuel economy regulations on the one hand, and the Obama Administration attempted to extend the federal program for regulating GHGs on the other, the

[28]Based on interview with AAM and Ford.

industry was willing to take the follow-up negotiations over the 2017–2025 model year standards. This is rigidly evident in the 2012 public hearing regarding the 2017–2025 fuel economy regulation held on January 2012. All US based manufacturers, industry networks such as AAM and Global Automakers, along with Japanese manufacturers, supported the harmonized fuel economy regulations between the federal and state levels (NHTSA and EPA 2012a, b, c). Such industry support came from the industry strategy to avoid adopting the multiple sources of regulations.

While supporting a harmonized fuel economy regulation up to 2025, the industry also asked for the addition of some flexibility measures. For instance, a 'mid-term review' provision was insisted on by the industry in order to make sure the government took a second look at the standards proposed for model years 2022–2025, so that the standard will remain appropriate and feasible, or to put it simply, the review allows to check whether technologies are advancing or not. If it were advancing faster, there would be a possibility to raise the standards; however, if it is not, it might have to lower the standards. Another example is 'off-cycle credits', in which automobile manufacturers would get credits for technologies that help to improve fuel economy. For instance, better air conditioning would generate credits for auto manufacturers to meet the standard.

In the 17th January 2012 public hearing, in Detroit, regarding the 2017–2025 fuel economy regulation, reasoning for the uncertainties of the future and the usefulness of the flexibility measures to help reduce CO_2 reduction from vehicles, all US based companies supported the need for the mid-term review and flexibility measures (NHTSA and EPA 2012a).

Finally, in contrast to the continued decline of the lobbying influence of the Big Three, the influence of foreign automakers increased over the fuel economy debate. While GM and Chrysler negotiated and got the US government's bailout, this led them to strongly oppose a very strong Obama administration priority. Japanese manufacturers, such as Honda and Nissan, supported the increase of CAFE regulation, and in return, "as international automakers increased their investments in the US, and therefore made contributions to the US, more and more members of Congress both in the House and the Senate supported us, and maybe became less supportive of the Big Three".[29]

In summary, the role of California and the new leadership of the Obama administration, as well as its newly appointed EPA, changed the dynamics towards the regulatory convergence of US fuel economy regulation up to the level of European and Japanese standards. This section argued that the competitiveness issues had been the prime driving force to push for such standards to be legalized in the US. Therefore, we can reasonably conclude that the California effect not only occurred within the US when California standards were pushed to raise the federal standards, but also internationally, where stringent European and Japanese standards influenced the construction of the new US CAFE regulation for 2025.

[29]Interview with JAMA US.

5.7 Summary

This chapter looked at a construction of the US fuel economy regulations. Table 5.4 describes the characteristics of the US fuel economy regulation.

On the contrary to the EU, energy-savings, rather than concerns for climate change issue, was the fundamental motivation in the United States. The US CAFE regulations were introduced in reaction to the oil crisis in 1970s. Nevertheless, the CAFE regulations were stagnated partly because of the cheap gasoline price during 1980s and 1990s. The CAFE regulations again improved in the late 2000s due to a significant rise of gasoline price, which motivated the policy makers to improve the regulation from a viewpoint of energy security. Furthermore, the subsequent Iraqi war, since 2003, and the US foreign policy on the Middle East fostered discussions among neoconservatives that support reduction of US oil independence. While energy-saving motive is the central driving force to pushing the CAFE regulation, climate change issue is becoming another motive, as the US EPA now joins the decision-making process with its authority to regulate car CO_2 emissions conferred by the Clean Air Act. This means that the US could introduce stringent CAFE regulation in the future motivated by a concern for climate change issue.

Even though the US was the first country in the world to introduce fuel economy regulation, primarily because of the strong industry lobby against an increase of the standard as well as the cheap oil price and subsequent lack of need to improve the standard, its standard has been stagnant for more than 20 years. Thus, business had been acted as laagered to stop any increase of CAFE regulations. The obvious side

Table 5.4 The characteristics of U.S. fuel economy regulation

Key factors	Description
Motive	Driven by the energy security concerns
Competitiveness issues	US auto industry gained much profits from the domestic market but later it lost its share in both domestic market and hence lost competitiveness. Fuel economy regulation aimed for 2025 aimed to recover the competitiveness
Decision-making process	Multiple sources of regulations. At the federal level, it is originally set by NHTSA, but EPA also gained authority to regulate under the Clean Air Act. At the state level, California introduced its own standards
Key actors	California played a critical role in pushing the CAFE forward. Industry lobbied to stop any increase of CAFE; but stagnant CAFE regulation in turn weekend competitiveness of auto industry. Environmental NGOs played a moderate role
Critical juncture	In 2007 when Supreme Court decided CO_2 as a pollutant. In reaction, California introduced own regulation, that in turn changed the behaviour of the industry to support improvement of CAFE regulations

Source Created by author

effect of stagnant fuel economy regulation is that the stagnant regulation actually weakened the US automobile industry's competitiveness over that of the Japanese industry. While the US automobile industry has been enjoying most of its SUV sales within the domestic market, this share of sales was gradually taken over by the Japanese automobile industry. This is especially evident in times of high gasoline prices, when consumers prefer more fuel-efficient cars.

Significant changes were brought forward in 2007 with the Supreme Court decision that the EPA has the authority to regulate car CO_2 emissions under the authority given to it by the Clean Air Act. This historic turnover provided further incentives to the environmentally leading state of California, which had originally adopted its own vehicle emissions standards in 2002. With the Supreme Court decision, California was granted the authority to implement its own GHG emission reduction standards for new passenger cars by the newly appointed EPA under the Obama administration. Because the industry feared California taking any further action that might potentially introduce its standards, and concerned like-minded states would follow California's standards, California's emission standards, in turn, changed the industry position to support a uniform regulation at both the state and federal level. Thus, industry wanted to avoid multiple sources of potential regulations, both at the federal level under the lead of the EPA (which can regulate CO_2 from cars as an 'air pollutant'), and the state level under the lead of California (which can go further ahead than federal level standards). Hence, California was the critical norm entrepreneur by leading the stringent emission standards, which gradually spilt over to the federal level. As a result, the Energy Independence and Security Act, which for the first time in 20 years raised the CAFE standards, was adopted in 2007.

With the emergence of the Obama administration in 2009, not only energy security issues, but also competitiveness issues of the US automobile industry

Table 5.5 The milestone of US fuel economy regulations

Year/Event	Milestones of US fuel economy regulations	Factors that influenced US fuel economy regulations
1990s	• 1990: California's LEV and ZEV regulations	• Business lobby to block any increase of CAFE; resulted in stagnation
2000s	• 2002: California passed 'Pavely Law' that set '323 g/km by 2009'; '205 g/km by 2012' and '172 g/km by 2016'	• 2007: Supreme Court decision of 'CO_2 as air pollutant'
	• 2007: Energy Independence and Security Act mandated '172 g/km by 2020'	
	• 2009: President Obama proposed '172 g/km by 2016'	
2010s	• 2011: President Obama proposed maximum '99 g/km by 2025 target'	
	• 2011: '103 g/km' by 2025 agreed	

Source Created by author

became the subject of raising the CAFE regulation. Perhaps the most fundamental concern was the bailout of the US automobile industry and its shrinking share within the domestic market. According to the interview conducted with policy officer at the US EPA, the US government concerned that if the US does not adopt as stringent fuel economy regulations as Europe and Japan, the US based automobile industry and related jobs will be more heavily damaged than any raise in the CAFE regulation. Consequently, the newly appointed EPA under the Obama administration proposed the stringent fuel economy regulations up to the level of the Japanese and European regulations; and it resulted in the regulatory convergence between Europe, Japan and the US. Thus, the competitiveness issues in the US pushed for the stringent CO_2 standards for cars in the absence of any legally binding national laws to limit CO_2. This suggests that, like the car industry, other US private sectors that lost competitiveness due to stagnant environmental standards in the US would potentially be motived to support stringent environmental standards to enhance its competitiveness.

Table 5.5 describes the milestone of US fuel economy regulations.

References

Bamberger R (2002) Automobile and light truck fuel economy: the CAFE standards. Washington DC

Bang G et al (2005) Future US climate policy: international re-engagement? International studies. Perspective 6(3):285–303

Boyle K (1998) The UAW and the Heyday of American Liberalism, 1945–1968. Cornell University Press, Ithaca

Cannon, J. Z. (1998). Memorandum: EPA's Authority to Regulate Pollutants Emitted by Electric Power Generation Sources, Washington DC

Cardell NS, Dunbar FC (1980) Measuring the societal impacts of automobile downsizing. Transp Res.-A 14A:423–434

CNN (2011) Cars sold in US could get 54.5 mpg by 2025

Cooney S, Yacobucci BD (2005) US automotive industry: policy overview and recent history—CRS report for congress

Cornell University Law School (2007) Massachusetts v. EPA (No. 05-1120). http://www.law.cornell.edu/supct/html/05-1120.ZO.html. Accessed 12 Mar 2013

Crandall R, Graham J (1989) The effect of fuel economy standards on automobile safety. J Law Econ XXXIII:97–118

Crandall R et al (1986) Regulating the automobile. Regulating the Automobile, Washington DC

EPA (n.d.) Clean Air Act

EPA (2003) Federal register vol 68, No 173

EPA (2008) 73 Federal register 44354: regulating greenhouse gas emissions under the Clean Air Act, Washington DC

EPA (2009a) 74 Federal register: proposed endangerment and cause or contribute findings for greenhouse gases under section 202(a) of the Clean Air Act, Washington DC

EPA (2009b) EPA will propose historic greenhouse gas emissions standards for light-duty vehicles, Washington DC

EPA (2010) Interim joint technical assessment report: light-duty vehicle greenhouse gas emission standards and corporate average fuel economy standards for model years 2017–2025. http://www.epa.gov/oms/climate/regulations/ldv-ghg-tar.pdf. Accessed 26 Mar 2012
EPA (2011) NHTSA, in coordination with California, announce plans to propose greenhouse gas and fuel economy standards for passenger cars and light trucks, Washington DC
EPA (2012) Federal register vol 77, No 199, Washington DC
EPA and NHTSA (2011) 2017–2025 Model year light-duty vehicle GHG emissions and cafe standards: supplemental notice of intent, Washington DC
Fabricant RE (2003) EPA's authority to impose mandatory controls to address climate change under the Clean Air Act
Ford Motor Company (2007) Annual report 2007. Michigan
Ford Motor Company (2008) Annual report 2008. Michigan
Ford Motor Company (2009) Annual report 2009. Michigan
Ford Motor Company (2010) Annual report 2010. Michigan
Freedman C, Blair A (2010) Seeds of destruction : the decline and fall of the US car industry. Econ Labour Relat Rev 21(1):105–125
Gartman D (2004) Three ages of the automobile: the cultural logics of the car. Theory Cult Soc 21 (4/5):169–195
Gerald D, Lave L (2003) The economics of CAFE reconsidered: a response to CAFE critics and a case for fuel economy standards, Washington DC
General Motors (2007) Annual report 2007. Detroit
General Motors (2010) Annual report 2010. Detroit
GTZ (2009) International fuel prices 2009 6th edn
Hove S et al (2002) The oil industry and climate change: strategies and ethical dilemmas. Clim Policy 2:3–18
ICCT (2011) Global passenger car fuel economy and/or greenhouse gas emissions standards—May 2011 update, Washington DC
ICCT (2014) global passenger car fuel economy and/or greenhouse gas emissions standards—May 2014 update, Washington DC
IEA (2009a) Transport, energy and CO_2: moving towards sustainability. Paris
IEA (2009b) CO_2 emissions from fuel combustion highlights—2010 edition. Outlook, Paris
Iguchi M, Hillman K (2012) The development of fuel economy regulation for passenger cars in Japan. In: Nilsson M et al (ed) Paving the road to sustainable transport governance and innovation in low-carbon vehicles. Routledge, London, pp 57–69
Inoue H (2008) Beikoku Chikyu Ondanka Taisaku no Doukou (The trend of US climate policies). Environ Law J 77–90
International Center for Technology Assessment (1999) Petition for rulemaking and collateral relief seeking the regulation of greenhouse gas emissions from new motor vehicles under the 202 of the Clean Air Act, Washington DC
King B (2012) Public Hearing before the NHTSA and the EPA regarding to 2017 and later model year light-duty vehicle greenhouse gas emissions and corporate average fuel economy standards. Detroit
Levy D (2005) Business and the evolution of the climate regime: the dynamics of corporate strategies. In: Levy DL, Newell PJ (eds) The business of global environmental governance. MIT Press, Cambridge
Levy D, Rothenberg S (1999) Corporate strategy and climate change. In: heterogeneity and change in the global automobile industry, Discussion Paper E-99-13, Kennedy School of Government, Harvard University, pp 1–27
Library of Congress (2006) Bill text 109th congress (2005–2006) S.3694.IS
McCarthy JE (2009) Cars and climate: what can EPA do to control greenhouse gases from mobile sources? Washington DC
McCarthy JE, Yacobucci BD (2013) Cars, trucks, and climate: EPA regulation of greenhouse gases from mobile sources, Washington DC

Meltz R (2007) The Supreme Court's climate change decision: Massachusetts v. EPA, Washington DC
Mikler J (2010) Business as usual or apocalypse now? Regulating the car industry's carbon emissions. In: Third biennial conference of the european consortium on political research standing group on regulatory governance, Dublin
National Academy of Science (2001) Effectiveness and impact of corporate average fuel standards, Washington DC
Newell P, Paterson M (1998) A climate for business: global warming, the state and capital. Rev Int Polit Econ 5(4):603–679
NHTSA (2004) Automotive fuel economy program annual update calendar year 2003
NHTSA and EPA (2012a) Transcript of the Detroit public hearing, EPA/NHTSA proposal standards for greenhouse gases and fuel economy hearing, Detroit
NHTSA and EPA (2012b) Transcript of the Philadelphia public hearing, EPA/NHTSA proposal standards for greenhouse gases and fuel economy hearing, Philadelphia
NHTSA and EPA (2012c) Transcript of the San Francisco public hearing, EPA/NHTSA proposal standards for greenhouse gases and fuel economy hearing, San Francisco
Nivola P, Crandall R (1995) The extra mile, Washington DC
Nordhaus M, Shellenberger T (2005) The death of environmentalism. Grist magazine
OICA (2007) World motor vehicle production OICA correspondents survery, Paris
OICA (2008) World motor vehicle production OICA correspondents survery, Paris
OICA. (2009) World motor vehicle production OICA correspondents survey. Paris
OICA (2010) World motor vehicle production OICA correspondents survey, Paris
Opernsecrets.org (n.d.) Annual lobbying on automotive. https://www.opensecrets.org/lobby/indusclient.php?id=M02&year=a. Accessed 12 Aug 2014
Paterson M (2007) Automobile politics. Cambridge University Press, Cambridge
Porter G, Brown JW (1996) Global environmental politics. Westview Press, Boulder
Schreurs MA (2003) Environmental politics in Japan, Germany and the United States. Cambridge University Press, Cambridge
Set America Free (n.d.) Set America free coalition members
Set America Free (2004) An open letter to the American people
Shin D (1990) The costs and benefits of federally mandated policies to promote energy conservation: the case of the automobile efficiency standard 1990, Washington DC
Source Watch (n.d.) Global climate coalition. http://www.sourcewatch.org/index.php?title=Global_Climate_Coalition. Accessed 29 Mar 2012
UAW (2012) Improved vehicle fuel economy to create 20,000 Michigan jobs
US Energy Information Administration (2013) Gasoline and diesel fuel update, Washington DC
US Department of Energy (2002) Freedom CAR partnership plan
US Supreme Court (2006) Transcripts of oral argument on Massachusetts et al. Petitioners v. Environmental Protection Agency et al, Washington DC
US Supreme Court (2007) Massachusetts et al v. Environmental Protection Agency et al. Certiorari to the United States court of appeals for the district of Columbia circuit, Washington DC
Vogel D (1997) Trading up and governing across: transnational governance and environmental protection. J Eur Pub Policy 4(4):556–571
White House (2007) President Bush Signs H.R. 6, the energy independence and security Act of 2007. http://georgewbush-whitehouse.archives.gov/news/releases/2007/12/20071219-6.html. Accessed 29 Mar 2012
White House (2009a) Obama administration national fuel efficiency policy: good for consumers, good for the economy and good for the country. http://www.whitehouse.gov/the_press_office/Fact-Sheet-and-Particpants-at-Todays-Rose-Garden-Event/. Accessed 29 Mar 2012
White House (2009b) President Obama announces national fuel efficiency policy. http://www.whitehouse.gov/the_press_office/President-Obama-Announces-National-Fuel-Efficiency-Policy/. Accessed 26 Mar 2012
Yacobucci BD, Bamberger R (2007) Automobile and light truck fuel economy: the CAFE standards, Washington DC

Chapter 6
Comparative Assessment

Abstract This chapter compares and contrasts fuel economy regulations in Europe, Japan and the US on the grounds of: (1) motive of the regulation in each studied country/region; (2) importance of competitiveness issue to enhance the regulations; (3) decision-making process and the roles of non-stare actors, including business and NGOs; and (4) the year of critical juncture that led to the enhancements of fuel economy regulations and the regulatory convergence. By assessing these points, it seeks policy implications that the regulatory convergence would place in considering the future climate governance for 2015 and beyond, namely, how the regulatory convergence between three regions may trigger the 'race to the top' of environmental regulations in emerging economies.

Keywords Motive of fuel economy regulations · Competitions issues and fuel economy regulations · Actors and fuel economy regulations · Decision-making process of fuel economy regulations · Critical junctures of fuel economy regulations · Race to the top of fuel economy regulations

6.1 Introduction

Throughout the Chaps. 3–5, the three empirical chapters examined developments of fuel economy regulations in Europe, Japan and the US, and focused on the emergence of the regulatory convergence by revealing who gets, what, when and how in the processes.

Based on these findings, this chapter compares and contracts these developments on the ground of: (1) motive of the regulation in each studied country/region; (2) importance of competitiveness issue to enhance the regulations; (3) decision-making process and the roles of non-stare actors, including business and NGOs; and (4) the year of critical juncture that led to the enhancements of fuel economy regulations and the regulatory convergence.

M. Iguchi, *Divergence and Convergence of Automobile Fuel Economy Regulations*,
DOI 10.1007/978-3-319-17500-3_6

By assessing these points, it draws implication to the future direction of fuel economy regulations beyond 2015 and beyond. Then, it further discusses the impacts of the regulatory convergence on the emerging economies, particularly China and India. It points out that the 'race to the top' of fuel economy regulations is emerging among major automobile manufacturing countries, regardless to the division of 'developed' or 'developing' countries in the climate negotiation.

6.2 Motive of the Regulation

This section compares motive of fuel economy regulations in each studied country/region. It points out that the regulations are increasingly driven by concerns for growing CO_2 emissions, and hence climate change, along with the competitiveness issue are becoming as the major driver to strengthening fuel economy regulations. To do so, it discusses in conjuncture with climate change negotiation being discussed at the UNFCCC, and how it would impact to the future regulatory convergence of fuel economy standards.

Among the all three case studies, both energy-savings and concern for climate change have been fundamental motivations of fuel economy regulations. However, weight of these rationales varies among the case studies. In Europe, while the energy saving has been concerned in order to reduce oil dependence and to promote the common market and enhance the competitiveness of its industry (European Commission 1991), climate change has been the main driver to promote its fuel economy regulations. This is obvious as the European regulation is expressed by 'grams of CO_2 per kilometre' while Japan and the US employs 'litters of gasoline per kilometre' or 'gallons of gasoline per miles'. This is not surprising, given that European fuel economy regulations emerged during 1990s, and discussions associating CO_2 emissions from cars were subject to the regulation given that the emissions was on the rise from the road transport sector.

In Japan, oil shocks and the state of high dependency on foreign oil triggered its first fuel economy regulations, hence energy saving was the fundamental motivation for the stringent regulations. At the same time, after the adoption of the Kyoto Protocol, reduction of CO_2 emissions from passenger cars became main motive for its fuel economy regulations. What is also interesting in Japanese fuel economy regulations is that while Japan does not participate to the second commitment period of the Kyoto Protocol (UN 2012), the 2020 target was primarily driven by the concern for the growing CO_2 emissions from the road transport sector.

The US CAFE regulation, which took place the earliest among the three, has been driven by the energy-savings and to reduce dependence on the foreign oil. Changes to the US fuel economy regulation was marked in 2007 when the US Supreme Court interpreted the 'CO_2 as an air-pollutant' and hence gave the EPA to regulate CO_2 emissions from passenger cars under the CAA (US Supreme Court 2006, 2007). Since then, California's regulation, in contrast to the federal CAFE regulation, is expressed on the grams of CO_2 per mile. The California's regulation then influenced improvements in the CAFE regulation at the federal level.

The question is, to what extent the UNFCCC regime influences the formulation of the fuel economy regulation? The Table 6.1 compares milestones of the UN-FCCC negotiations and climate policies in Europe, Japan and the US.

After the adoption of the UNFCCC in 1992, the EU took the lead in fuel economy regulations by proposing 120 g/km target by 2005. After the Kyoto Protocol was adopted in 1997,[1] the progress of fuel economy regulation was particularly advanced in Japan, which set 2010 target which required 22.8 % improvement compared to 1995 level with an introduction of the top runner method.

The improvements in the fuel economy regulations in Europe, Japan and the US were brought during 2000s. In Europe, the Commission proposed mandate reduction of '120 g/km (130 g/km) by 2012' and possibility to research towards '95 g/km by 2020' in 2007. So too, in Japan, '125 g/km by 2015' target was introduced in 2007. In the US, California set '323 g/km by 2009', '205 g/km by 2012' and '172 g/km by 2016' target under the Pavely law in 2002. More importantly, the former President Bush introduced 2007 Energy Independence and Security Act that mandated '172 g/km by 2020'. This federal level improvement of fuel economy regulation was further advanced by the President Obama in 2009, by setting '172 g/km by 2016' target. Hence, the year 2007 marked the critical junctures to fuel economy regulations in these three case studies.

The year 2007 was in fact the time when Parties to the UNFCCC agreed on the 'Bali Action Plan' at the COP13, that established "a comprehensive process to enable the full, effective and sustained implementation of the Convention through long-term cooperative action, now, up to and beyond 2012, in order to reach an agreed outcome and adopt a decision" (UN 2007, p. 3) towards the COP15. At the COP15 held in Copenhagen in 2009, Denmark, 'Copenhagen Accord'[2] was *taken note* by the COP. Countries submitted non-binding emission reduction pledges to this agreement as follows: the EU pledged 20–30 % reduction against 1990 level in accordance to the '2020 climate and energy package' (European Commission 2010); the US pledged 13 % against 2005 level (US Department State 2010); Japan pledged 25 % reduction against 1990 level (Japan Embassy in Germany 2010). Developing countries have also submitted their voluntary reduction targets: China pledged to reduce carbon intensity by 40–56 % compared to 2005; India pledged to reduce carbon intensity by 20–25 % compared to 2005; Brazil pledged 36.1–38.9 %

[1]The Kyoto Protocol, formally adopted at COP3 in Kyoto, set binding targets for 27 countries and European community to reduce an average of 5 % against 1990 levels over 2008–2012.

[2]Under the Copenhagen Accord, countries pledge their emission reduction target for 2020, agreeing that "deep cuts in global emissions are required according to science, and as document by the IPCC Fourth Assessment Report with a view to reduce global emissions so as to hold the increase in global temperature below 2 °C" (UNFCCC 2009, p. 2) although the Accord is not legally binding in nature.

Table 6.1 Comparison of developments in climate mitigation policies in the UNFCCC, Europe, Japan and the US

Year	Milestones in the UNFCCC	Europe	Japan	US
1990s	**1992**: adoption of the UNFCCC	**1996**: adoption of limiting the 2 °C compare to pre-industrial level target		
	1997: adoption of the Kyoto Protocol	**1997**: 8 % reduction pledge under the Kyoto Protocol	**1997**: 6 % reduction pledge under the Kyoto Protocol	**1997**: 7 % reduction pledge under the Kyoto Protocol
2000s	**2007**: adoption of the Bali action plan	**2007**: EU set 20 (30) % reduction target by 2020 under the 2020 climate and energy package	**2009**: 15 % reduction target against 2005 level was proposed by Prime Minister Aso (equivalent to 8 % reduction against 1990 level)	**2000**: withdrawal from the Kyoto Protocol **2002**: California set a target of 25 % compared to 1990 levels by 2020
	2009: Copenhagen Accord was taken note by the COP		**2009**: 25 % reduction target against 1990 level was proposed by Prime Minister Hatoyama	**2009**: Whitehouse proposed 17 % reduction against 2005 level by 2020
2010s	**2011**: adoption of Durban Platform	**2010**: pledged 20 (30)% reduction against 1990 level to the Copenhagen Accord	**2010**: pledged 25 % reduction against 1990 level to the Copenhagen Accord	**2010**: pledged 17 % reduction against 2005 level to the Copenhagen Accord
	2012: adoption of the Doha Amendment to the Kyoto Protocol	**2011**: the Council considers 80–95 % reduction target by 2050	**2013**: adopted new target of 3.8 % reduction against 2005 level by 2020	

Source created by author

reduction compared to business as usual; and South Africa pledged 34 % reduction compared to business as usual (UNFCCC 2010).

At the same time when the international society move to discusses climate mitigation target for 2020,[3] so did fuel economy regulation for 2020 progressively

[3]For example, in 2011, Parties to the UNFCCC decided to launch a process to "develop a protocol, another legal instrument or an agreed outcome with legal force under the Convention applicable to all Parties" (UNFCCC 2011, p. 2), through a subsidiary body under the Durban Platform for Enhanced Action, which shall "complete its work as early as possible but no later than 2015 in order to adopt this protocol, legal instrument or agreed outcome with legal force at the twenty-first

Table 6.2 Comparison of rationales of fuel economy regulations between Europe, Japan and the US

	Europe	Japan	US
Motive	Climate driven	Mix of energy-saving and climate driven	Energy-saving driven
Influence of climate change issue on fuel economy regulation	High, fuel economy regulation as a part of climate mitigation policy	High, the adoption of the Kyoto Protocol and after as one of the major rationales	Low at the federal level, high at the state level (California)

Source created by author

evolved: in Europe, the Commission is reviewing to propose mandate target of '95 g/km by 2020' in 2011; Japan started to set '105 g/km by 2020' target in 2010 and introduced '105 g/km by 2020' target in 2011; and in the US, '103 g/km by 2025' target was agreed in 2011.

To summarize, although it is overemphasizing to say that the developments of climate change agreements directly influenced the evolutions of fuel economy regulations, it is clear that the timing of the regulations in each country/region have been developing accompanying with the progress of the international climate change agreements under the UNFCCC. Table 6.2 compares climate change as a rationale of fuel economy regulations in Europe, Japan and the US. Given that official documents and interviews conducted to various stakeholders confirm that the fuel economy regulations for 2015 and 2020 in Europe and Japan were to a large extent motivated to reduce CO_2 emissions from road transport sector, and considering that California's climate-driven regulations and its impacts upon the federal level CAFE regulation, concerns on climate change issue would continue to play a critical role in deciding the target for beyond 2020.

Having said that, another rationale that pushed improvements of fuel economy regulations is to enhance competitiveness of automobile manufactures in each country. The next section explains how the enhancements of competitiveness of the manufactures triggered the regulatory convergence.

6.3 Importance of Competitiveness Issue in Enhancing Regulations

I have pointed out that the EU's environmental policy emerged out of the concerns among member states that diverse environmental standards could result in trade barriers and competitive issues in the common market (Johnson and Corcelle 1989).

(Footnote 3 continued)
session of the Conference of the Parties and for it to come into effect and be implemented from 2010" (UNFCCC 2011, p. 2).

Business competitions in European automobile industry arose when Germany proposed to improve the fuel economy of new cars to an average of 5 L/100 km (120 g/km) by 2005 in 1994. This means that German proposal in 1994 was made in order to take first mover's advantage critical to determine the remainder of the target: it pushed for a percentage reduction target of 75 % against 1990 levels by 2005 and the introduction of a weight-based system that would benefit to German automobile industry over the French industry. On the contrary, France insisted on regulation based on absolute figures expressed by 'CO_2-g/km' with incentives for those who achieved the target, such as financial benefits, while those who exceed the target would pay a fine (The ENDS Report 1997). Although German preferences have reflected to a great extent in the final agreement, however, France, which produces fuel efficient cars, supported and pushed the stringent 95 g/km by 2020 target. Consequently, the European fuel economy regulations have been progressed by conflicts over business interests between Germany and France, both of which are motivated to increase the competitiveness of its automobile industry within the European market.

Regulations in the EU and to the extent the US influenced construction of the Japanese fuel economy regulations. As one of the major car manufacturing countries, Japan exports great numbers of cars. The US is Japan's biggest car export market with 30.7 % of the total numbers of four-wheel cars exported in 2008, while Europe is the second biggest market with a share of 23.6 per cent. Although Asia is growing to be among the biggest export markets, its share is not yet as big as those of the US or the EU. In Europe, a voluntary target was agreed between JAMA and the European Commission, aiming for the target of 140 g/km by 2009. As a result, this target may have worked as one of the driving forces that pushed improvement in the Japanese industry. If we compare timings with the European fuel eocnomy regulation, it is obvious that Japanese regulation are introduced as almost the same timing the European new regulation has proposed. For example, when the discussion of legalization of European 120 g/km target emerged in 2007 and adopted in 2009, which is generally considered as the political target rather than the target based on technological potentials, the Japanese target for 2015 (125 g/km) is also adopted in 2009. Furthermore, as the European 95 g/km by 2020 target was also embedded in the 2009 legislation, Japan began to set 105 g/km by 2020 target right after 2010.

In the US, business had been acted as laagered to stop any increase of CAFE regulations. Nonetheless, the stagnant CAFE actually weakened the US automobile industry's competitiveness over that of the Japanese industry. While the US automobile industry has been enjoying most of its SUV sales within the domestic market, its share was gradually taken over by the Japanese automobile industry and revealed the lack of competitiveness of the US automobile industry. According to interviews conducted with a senior policy officer at the US EPA and the vice president of the environment department of AAM, the competitiveness concern was the primal factor in formulating the 2017–2025 CAFE regulation.

These series of events, dynamics and trends suggest that automobile manufactures in each country are now competing over the technological innovations on

Table 6.3 Comparisons of competitiveness issues in fuel economy regulation in studies cases

	Europe	Japan	US
Importance of competitiveness issue against foreign manufactures	Primitive	Primitive	Important after the bailout of auto industry
Motivation to enhance competiveness issue	In order to enhance the competitiveness within the European market	In order to stay competitiveness in foreign markets	In order to enhance competitiveness within domestic market

Source created by author

low-carbon vehicles. This in turn implies that a country with higher low-carbon technological standard may impose stricter automotive fuel economy standards, so that it could further encourage low-carbon technological innovations as well as to competences on its domestic automobile manufactures against other manufactures based in low fuel economy regulations. If such manufactures wish to continue to export their products, they may have to comply with the stricter standards and therefore they are likely to support stricter standards in their home country (Volgel 1997). Table 6.3 summarises the comparisons of competitiveness issues in fuel economy regulations in studied cases.

Having touched upon the importance of concerns on climate change issue and rationales to enhance competitiveness of automotive manufactures that pushed the regulatory convergence of fuel economy regulations, the next section compares and contrasts how the actual decision-makings were taken place.

6.4 Decision-Making Process

To begin with, it is critical to point out that Varieties of Capitalism approach fails to grasp the current decision-making structures in Europe and Japan, and to an extent, in the US. Firstly, although existing literatures show that both the EU and Japan shares a characteristic of the Coordinated Market Economy, in which decision-making processes are based on co-regulations and corporatism (Mikler 2007); however, this book revealed that there are significant differences exist between the two. It is true that in both regulations, the state and corporations negotiate policy by consultations, although the members of corporations do not have significant involvement in the process. Nevertheless, if the very concept of co-regulation and corporatism suggest both state and corporations form social partners to create particular policies through cooperation, consultation, negotiation and compromise (Wiarda 1996), the European regulations can be sketched as the clash of divergent interests among member states, while Japanese regulations is characterised as rather harmonious government-in-lead regulation.

In the case of Europe, the fuel economy regulations are based on a supranational codecision-making process between the Commission, the Council of the EU and the European Parliament. An authority to propose new regulatory standards is conferred to the Commission, and the proposal is discussed at both the Council of Ministers and the Parliament in parallel. Each institution represents different interests: while the Commission represents the interest of the EU as a whole rather than each member state; the Council of the EU represents interests of each country; and the Parliament represents both interests of the industry and NGOs. This means automobile industry in each member states has multiple channels to exert its influence—through consultation with the Commission, through Ministers in the Council, and through Industry Committee in the Parliament. As a result, industry successfully weakened the original Commission proposal of 120 g/km by 2012, to the level of 130 g/km by 2015.

In Japan, the development of fuel economy regulation was achieved through 'co-regulation', where government, industry, and academia participated in the standard setting process. The industry took part in the decision making process through the central industry network for the Japanese car industry, JAMA. After an agreement was reached in these meetings, it was passed on to the Council for examination. Their role in this process is to discuss the appropriateness and feasibility of the new standard, by examining the Japanese car industry's technological potential. These series of decision-making procedures enabled incremental improvements of the Japanese fuel economy regulations, in contrast to the EU regulation that took 13 years to legalize its fuel economy regulations.

In the US, its CAFE regulations until the late 2007 can be characterised as the Liberal Market Economy. The government and business does not coordinate their position in the decision-making processes, but rather, a strong business lobbying to the federal government influenced the stagnated improvements in CAFE regulations. In other words, the stagnated CAFE were caused by business influence over the Congress and its consequence of low Congressional appropriation on the NHTSA.

However, Liberal Market Economy model does not explain the recent improvements of the CAFE standards nor it captures the changing nature of its business-government relationships. With the California's own legislation to limit CO_2 emission from cars, and the leadership of the White House in proposing stringent fuel economy regulation and the resulted EPA's involvement to the regulatory process indicates the emergence of more divergent, plural decision making process in the US where a more stringent regulation could be proposed at both state and federal levels. Also, critical change was brought the decline of the US automotive manufactures' share in the domestic market, which has been taken over by Japanese manufactures. As the share of the US automotive manufactures decline, many interviewees observed the relative decline of their influence over the Congress. Furthermore, in April 2009, when the EPA proposed the 103 g/km target by concluding the target could be achieved cost-effectively based on the calculation, the related stakeholders—EPA, NHTSA, California, automobile manufactures, the UAW and environmental NGOs—discussed this proposals and reached an

Table 6.4 Comparison of decision-making processes between Europe, Japan and the US

	EU	Japan	US
Characteristics	Co-decision procedure between supranational institutions	Co-regulation	Multiple sources of regulations
Key governmental actors	European Commission, European Parliament and European Council	MLIT and METI	NHTSA, EPA, California, White House, Congress and the Supreme Court

Source created by author

agreement among them. Considering these changing dynamics in the US fuel economy regulations, further reconsideration of the US Liberal Market Economy model may be needed in order to observe the future trend of the US fuel economy target for 2015 and beyond.

To summarize, this section revealed the differences of European and Japanese Coordinated Market Economy, and pointed out the changing nature of the US Liberal Market Economy model. Table 6.4 compares decision-making processes between the three case studies. The next section examines the critical junctures that led to the regulatory convergence and the role of societal actors that influenced the processes in each country/region.

6.5 The Critical Juncture and the Role of Non-state Actors

Different actors contributed to the critical change of fuel economy regulations in each country/region that led to the regulatory convergence. Non-state actors include environmental NGOs, business actors, local governments and scientists/experts.

In the EU fuel economy regulations, the critical junctures were brought by environmental NGOs, that acted as 'norm entrepreneurs' by revealing the industry was way below the voluntary agreement standards. They started to publicize that automobile manufactures were way below from the standards set by voluntary agreement with the European Commission. In turn, this led the Commission to propose a mandatory target of 120 g/km by 2015 with a scope of 95 g/km target by 2020 in 2007 and hence created a path for significant change in the EU fuel economy regulations. Furthermore, in the subsequent discussions in the European Parliament, environmental NGOs were also actively engaged to support 95 g/km by 2020 target, with a view for a further reduction of 70 g/km target by 2025.

In contrary, ACEA was opposing against the implementation of the regulation by 2012, arguing that the timetable should be 2015 due to the lack of planning certainty. Also, it claimed that the target itself should be 135 g/km instead of 120 g/km, arguing that 15 g/km should be achieved by using the complementary approach, namely, an introduction of bio-duels, diffusion of eco-driving and so forth. While French and Italian manufactures had different opinions, it is notable that the position of ACEA was very much dominated by the German manufactures.

Scientists and experts also played important role in the EU fuel economy regulations by proposing indicative target for 2020. In contrary to the 120 g/km target can be described more as a political goal, which was originally proposed by the European Commission in 1995 and initiated to legalize in 2007, a 95 g/km targeted is based on calculations, which originates in the scientists' indicate numbers ranging from 85 to 105 g/km on preliminary assessments of possible targets for 2020. Although it is assumed that the final target of a 95 g/km was picked as a result of political bargaining among various actors at various levels, the target is within the range of indicative target proposed by scientists.

In Japanese fuel economy regulations, NGOs did not play a role. Rather, the whole process was led by the Ministries. In this process, business actors involved in the decision-making process, and interestingly, Japanese car manufactures have already achieved the 2010 target for gasoline cars by 2007. As a result, 125 g/km target was introduced in 2007 for gasoline cars with the target year 2015, which is as stringent as the European target for 2015. Also important in this process is the role of the council that comprises of experts (e.g. academics). As the council can accept or reject the proposed targeted based on technological potentials, they are critical to the target setting.

In the US, where fuel economy regulation was stagnant for more than 20 years mainly due to the strong industry lobby against an increase of the standard, the critical juncture was brought by the Supreme Court decision that defined that 'CO_2 is an air pollutant' and thus the EPA was granted an authority to regulate CO_2 emissions from cars under its authority conferred by Air Pollution Act. This in turn led the state of California to enact its own air pollution standards for passenger vehicles, which have been imploring the EPA to regulate CO_2 emissions from cars under the Clean Air Act. California's adoption of its own standards for passenger vehicles triggered the 'California effect', which encouraged the federal government to harmonize its standard with the state of California. Consequently, California played key role as a norm entrepreneur in the US fuel economy regulations. The California's adoption of its own fuel economy regulation influenced the industry behaviour and thus changed the whole dynamics in the US CAFE regulation. As a result of California's adoption, the White House took leadership and proposed for an ambitious improvement of CAFE regulation. To do so, it appointed the EPA and NHTSA to conduct a study the feasibility of the proposed target. Consequently, the EPA and the NHTSA came down to a conclusion with a range of 99–131 g/km by 2025 and thus provided a scientific basis of the regulations.

The US environmental NGOs played a moderate role. Perhaps its notable activity is the submission of a petition in 1999, to request the EPA to regulate GHG emissions from motor vehicles under the CAA. Although this petition did not result in a success as the EPA denied it, they prompted a lawsuit against the EPA (as known as (*Massachusetts v. EPA 549 U.S. 497*) together with 12 states including Massachusetts and California, and several cities including New York during 2006

Table 6.5 Comparisons of critical juncture and key societal actors in fuel economy regulations in Europe, Japan and the US

	EU	Japan	US
Year of critical juncture	2007	2007	2007
Critical juncture	When the Commission proposed a mandatory 120 g/km target with a view to 95 g/km target	When manufactures achieved the 2010 target by 2007, and 2015 target introduced which is as stringent as the EU's standard	When the Supreme Court decision was made and California set its own regulation; and the CAFE regulation improved for the first time in 20 years
Critical players	NGOs and European Commission	Government (MLIT and METI) and JAMA	California, White House, the Supreme Court, EPA, NHTSA and NGOs
Business role	Business conflicts between France and Germany have been critical	Rather cooperative	Lobbied but later accepted CAFE increase
NGOs role	High, publicized industry's underachieves of a voluntary agreement that led to the Commission to propose the regulations	NGOs did not involved in the decision-making process	Moderate, engaged into a law-suit together with California and like-minded states
Role of scientists	Moderate, proposed indicative target for 2020	High, examination by the Council critical in deciding the target in both 2015 and beyond	Moderate, EPA and NHTSA proposed ranges of target for 2020

Source created by author

and 2007 that led to the Supreme Court decision in 2007 and resulted in California's own standard. Table 6.5 compares critical junctures and key actors in fuel economy regulations in Europe, Japan and the US.

6.6 Implications for 2015 and Beyond

The last section revealed more detailed characteristics of fuel economy regulations in Europe, Japan, and the US based on the analysis of political dynamics in each country and the region. Based on the comparison, this section provides practical implications that are drawn from case studies. It firstly explains whether the same logic of the regulatory convergence is (or could) happening for heavy-duty vehicles or not. It then point out the practical implication to improve global climate governance, especially regarding discussions associating with the 'race to the top' or 'race to the bottom', and concludes the discussions by proposing future research questions.

Implication to Regulatory Convergence for Heavy Duty Vehicles

The first practical implication of the regulatory convergence of car fuel economy is the question regarding whether the regulatory convergence is applicable to *heavy duty vehicle* fuel economy. Heavy duty vehicles, such as trucks are one of the biggest sources of CO_2 emissions in road transport sector in each studied countries/ region. For instance, heavy duty vehicles account for about six per cent of Japanese and European total CO_2 emissions (European Commission 2013; MLIT n.d.).

Furthermore, Europe, Japan and the US are also major producers of heavy duty vehicles. Table 6.6 compares productions of heavy duty vehicles between Europe, Japan and the US in 2011 and 2012. Among these three, Japanese manufactures dominate global heavy duty vehicle production in 2010, headed by Isuzu (12.9 %), and followed by Toyota (5.8 %) and Mitsubishi (0.09 %). European manufactures, such as Daimler (8.7 %) and Volvo (4.8 %) are also major producers of heavy duty vehicles. Although smallest among three, the US is a relatively large heavy duty vehicle producer, led by Ford (1.9 %), Chrysler (0.19 %) and GM (0.03 %) (OICA 2013).

This book specifically asked about the regulatory convergence of passenger car fuel economy because its primal interest lies in how car companies that operate in different countries could be the driver to push for stringent environmental policy-making. Unlike passenger cars that are designed to transport passengers, heavy duty vehicles such as trucks are designed to transport commercial goods. Commercial vehicle are said to be more sensitive to fuel economy than passenger vehicles because there has always been an assumption that market should ensure optimum fuel efficiency and governmental policy interventions may not always be needed (IEA 2010, p. 31). Hence, fuel economy regulations for heavy trucks have not been developed concretely as compared to passenger car fuel economy regulations.

Table 6.6 Comparisons of heavy truck productions between Europe, Japan and the US, 2011–2012

Year/ Country	Europe	Japan	US	Total
2011	340,871 (8.5 %)	512,260 (12.7 %)	243,047 (6 %)	4007,480 (100 %)
2012	319,555 (8.5 %)	583,074 (15.5 %)	267,944 (7.1 %)	3743,510 (100 %)

Source OICA (2013)

Interestingly, in the case of Europe, there is no regulation set for heavy truck fuel economy, due to the high price of gasoline price that already contributes to reduce CO_2 emissions from this sector (IEA 2010, p. 34). It is only very recently that the Commission implemented technical studies to reduce CO_2 emissions from heavy duty vehicles due to the rising road freight traffic. It has just started consultation process to require a mandatory efficiency reporting by industry to develop its regulations (European Commission 2013). In contrary, Japan has been leading the fuel economy regulations for heavy duty vehicles, which was introduced in 2005 with the targets to achieve the fleet average of 7.09 km/L for trucks and 6.30 km/L for buses, to be effective from 2015 onwards.[4] Japan introduced such regulations prior to any other countries because CO_2 emissions from heavy duty vehicles are the second biggest sources of CO_2 emission from passenger cars within its transport sector (JAMA 2006).

As for the United States, too, a proposal to set standards for heavy vehicles was only made in 2010 and just adopted its final rule in 2011. However, the average fleet of fuel economy regulations set for vocational trucks is 273 g/mile, starting from 2014 onwards, marking the most stringent targets than in Japan or in Europe (EPA 2011).[5]

Hence, the regulatory convergence of fuel economy regulations for heavy duty vehicles is not happening yet, and would be difficult to happen following the same logic as the case from passenger cars due to two reasons: first, as pointed out, commercial vehicle are more sensitive to oil price than passenger cars, and therefore stringent regulations have not always been necessary because there has been an assumption that market would ensure optimum fuel efficiency of heavy duty vehicles; and secondly, the numbers of stakeholders that would be affected by regulations are larger than passenger car regulations. Since the main use of heavy duty vehicles are to transport commercial goods, the regulation would not only affect car manufactures but also any company that use transport service, transport industry itself, and consumers.

'Race to the Top', not 'Race to the Bottom'

I argue that stringent fuel economy regulation in major automobile manufacturing and importing nations, such as Europe, Japan and the US, would function as the international standard that emerging automobile manufacturing countries such as China and India would follow. Hence, the regulatory convergence is born out from regulatory competition among the major automobile manufacturing nations with the

[4]This can be translated into 369.6 g/km for trucks and 416 g/km for busses (transportpolicy.net n.d.).

[5]It is divided into following three categories: 373 g/mile for light heavy class (8501–19,500 lb); 225 g/mile for medium heavy class (19,501–33,000 lb); and 222 g/mile for heavy-heavy class (>33,001 lb).

rationale to enhance its competitiveness of the automobile industry. In turn, it would function as a 'structure' that influences the behaviours of each state and hence also function as the 'de facto standard' in the global automobile market.

I have mentioned that China is the biggest automotive producer in the world, producing almost as twice as much as Japan. Still, if we focus on global automobile production in terms of individual automotive manufacturers, Japanese manufacturers (Toyota, Honda, Nissan and Suzuki), European manufacturers (Volkswagen, PSA, Fiat and Renault), and U.S. manufacturers (General Motors, Ford and Chrysler) dominate the global automobile production market. This means that, even though China has become the biggest producer in the world, automobile manufacturers from Europe, Japan and the US. dominate most of the domestic production in China—for example, Chinese passenger car market is dominated by General Motors (17.8 %), Volkswagen (14.6 %), Hyundai-Kia (8 %), Nissan (7.5 %), Toyota (6.7 %), Honda (5.7 %), Ford (4.4 %), Chery (4.1 %), Geely (2.2 %) and others (29 %) (Business Insider 2010).

As a matter of fact, Chinese and Indian proposed fuel economy regulations are also converging with Europe, Japan and the US. Figure 6.1 indicates the converging trend of fuel economy regulations in Europe, Japan, the US, China and India.

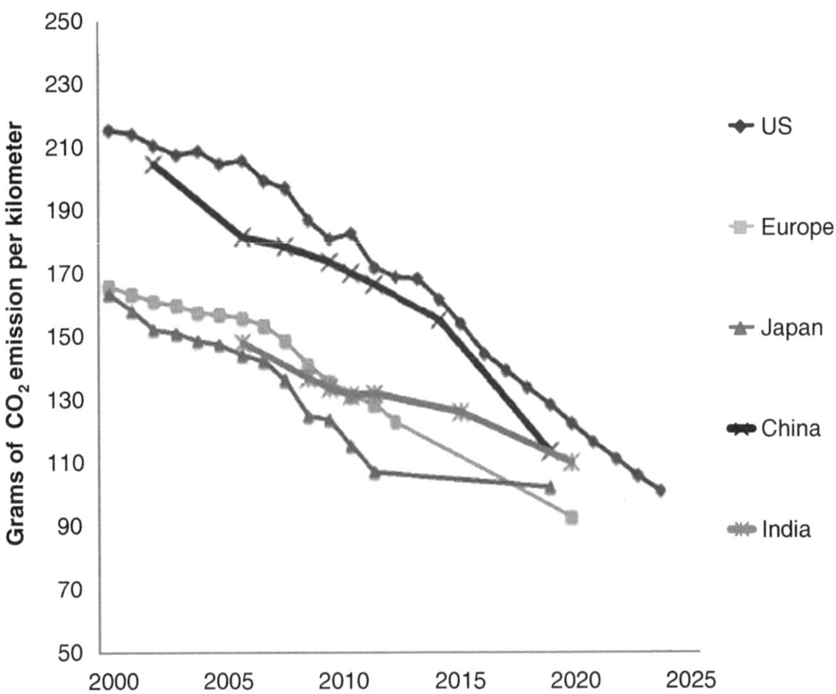

Fig. 6.1 Regulatory convergence between Europe, Japan, US, China and India. *Source* created by author based on ICCT (2014)

In China, the Ministry of Industry and Information Technology that is responsible for fuel economy regulations, issued a proposal of 117 g/km (5/100 km) as a 2020 target in the Energy-Saving and New Energy Vehicle Plan (The State Council of China 2012). One of the reasons for this introduction of a stringent fuel economy regulation could be said that while China is far behind of Japanese, European and perhaps the US competitors in terms of conventional vehicle technologies, it attempts to enhance its competitiveness by investing on new energy technologies (Kokko and Liu 2012).

In India, the transportation sector accounts for the second-largest contributor of the country's GHG emissions with a rapid growth due to the increasing number of vehicles (Bansal and Bandivadekar 2013). Furthermore, as India is the world's fourth-largest consumer of oil, with a high dependence on imported oil, the central government in consultation with the Bureau of Energy Efficiency under the Energy Conservation Act proposed fuel economy standards of 130 g/km for 2016 and 113 g/km for 2021 in January 2014 (Government of India 2014; ICCT 2014). This means that manufactures will have to improve the fuel efficiency at least 38 % by 2022, that will "catapult India into the league of select nations including the United States, Germany, Japan and China that strictly enforce norms with hard penalties foe violations" (The Economic Times 2014) and "car makers will have to make further investments in new technologies in order to achieve greater fuel efficiency" (Raj 2014).

This implication clearly demonstrates the 'race to the top' over regulatory competition and convergence, rather than 'race to the bottom'. The 'race to the top' of the regulatory convergence suggests countries compete over the stringent environmental standards for the sake of enhancing industry competitiveness. Consequently, it is reasonable to claim that what guides such regulatory competition is based on the normative consideration to stay competitive in the international automobile market. This in turn leads us to conclude that the higher the fuel economy standards in major automobile manufacturing regions, the more likely the standard automatically works as the global 'de-facto' standard in the world.

References

Bansal Gaurav, Bandivadekar Anup (2013) Overview of India's vehicle emission control program: past success and future prospects. ICCT, Washington, DC

Business Insider (2010) Beijing's new car restrictions are going to devastate the auto industry in 348 2011. http://articles.businessinsider.com/2010-12-28/markets/34930062006_1_auto-sales-beijing-municipal-commission-auto-industry. Accessed 7 Aug 2014

EPA (2011) NHTSA, in coordination with California, announce plans to propose greenhouse gas 955 and fuel economy standards for passenger cars and light trucks, Washington, DC

European Commission (1991) Communication from the commission to the council: a community 1081 strategy to limit carbon dioxide emissions and to improve energy efficiency. Brussels

European Commission (2010) Expression of willingness to be associated with the Copenhagen Accord and submission of the quantified economy-wide emissions reduction targets for 2020. European Commission, Brussels

European Commission (2013) Towards a strategy to address CO_2 emissions from heavy-duty vehicles. http://ec.europa.eu/clima/policies/transport/vehicles/heavy/index_en.htm. Accessed 4 Sept 2013
Government of India (2014) Gazette of India REGD. NO. L.-33004/99. Government of India, New Delhi
ICCT (2014) Global passenger car fuel economy and/or greenhouse gas emissions standards—May 2014 Update. Washington DC
IEA (2010) Transport energy efficiency: implementation of IEA recommendations since 2009 and next steps, Paris
JAMA (2006) Fuel economy target for heavy duty vehicles, Tokyo. http://www.359jama.or.jp/eco/weight/pdf/weight_mileage.pdf
Johnson SP, Corcelle G (1989) The environmental policy of the european communities. Graham & Trotman, London
Kokko A, Liu Y (2012) Governance of new energy vehicle technology in China. In: Nilsson M, Hillman K, Rickne A, Magnusson T (eds) Paving the road to sustainable transport governance and innovation in low-carbon vehicles. Routledge, London
Mikler J (2007) Varieties of capitalism and the auto industry's environmental initiatives: institutional explanations for firms' motivations. Bus Politics 9(1):1–38
OICA (2013) World Motor Vehicle Production by country 2011–2012, Paris
Raj A (2014) Norms for fuel efficient to put auto makers under pressure. http://www.livemint.com/Industry/XctQqjlxvTagxqh4ZzsD3K/Norms-for-fuel-efficiency-to-put-auto-makers-under-pressure.html. Accessed 16 Sept 2014
The Economic Times (2014) 14 % increase in mileage of passenger vehicle is mandatory now. http://articles.economictimes.indiatimes.com/2014-02-14/news/47336604_1_new-norms-fuel-efficiency-passenger-vehicles. Accessed 16 Sept 2014
The Embassy of Japan in Germany (2010) Note Verbale. The Embassy of Japan in Germany, Bonn
The ENDS Report (1997) Agreement on car efficiency heads for the rocks. London
The State Council of China 2012 Energy-Saving and new energy vehicle industrialization plan. http://www.gov.cn/zwgk/2012-07/09/content_2179032.htm. Accessed 7 Sept 2014
United Nations Framework Convention on Climate Change (UNFCCC) (2007) Report of the conference of parties on its fifteenth session, held in Bali from 3 to 15 Dec 2007. UNFCCC, Bali
United Nations Framework Convention on Climate Change (UNFCCC) (2009) Copenhagen Report. UNFCCCC, Copenhagen
United Nations Framework Convention on Climate Change (UNFCCC) (2010) Compilation of information on nationally appropriate mitigation actions to be implemented by parties not included in Annex I to the convention. UNFCCCC, Bonn
United Nations Framework Convention on Climate Change (UNFCCC) (2011) Report of the conference of the parties on its seventeenth session, held in Durban from 28 Nov to 11 Dec 2011. UNFCCC, Durban
United Nations (UN) (2012) Kyoto Protocol to the United Nations framework convention on climate change—Doha amendment to the Kyoto Protocol. UN, New York
US Department of State (2010) US submission. Office of the Special Envoy for Climate Change, United States Department of State, Washington, DC
US Supreme Court (2006) Transcripts of oral argument on Massachusetts, et al., petitioners v. Environmental Protection Agency, et al. Washington DC
US Supreme Court (2007) Massachusetts et al. v. Environmental Protection Agency et al. Certiorari to the United States court of appeals for the District of Columbia circuit. Washington DC
Vogel D (1997) Trading up and governing across: transnational governance and environmental protection. J Eur Publ Policy 4(4):556–571
Wiarda H (1996) Corporatism and comparative politics. M.E. Sharpe, New York

Chapter 7
Conclusion

Abstract This concluding chapter provides broader implications to theories, explores applicability of 'agency with and beyond the state' model to other sectors, and to climate governance as a whole, by answering research questions. Firstly, it describes how the proposed monograph made theoretical contributions to Varieties of Capitalism approach and Constructivist theory of International Relations. Then, it explains how this proposed monograph advanced the study of private environmental governance by pointing out the new model of 'agency with and beyond the state'. It argues that the model is applicable to other sectors that have following characteristics: (1) multinational corporations that are sensitive to environmental standards of export markets; (2) improvements of environmental quality of their products (such as higher energy efficiency or reducing hazardous substances) co-benefits to an enhancement of its competitiveness against their rivals.

Keywords Agency with and beyond the state · Private environmental governance · Business competition and environmental regulation · Beyond the varieties of capitalism · Regulatory convergence and environmental policy

7.1 Answering the Research Questions: Business Competition and Environmental Policy-Makings

This section answers research questions. It firstly tests hypothesis set in Introduction. It confirms that the second hypothesis, namely global competitive pressures, caused the regulatory convergence. Based on this finding, it then answers the research questions by explaining the mechanisms of competition over the regulatory convergence. The fundamental argument is that Japanese and the US fuel economy regulations are catching up with the European regulation because that seems to be the most cost-effective strategy.

Developing upon the constructivist theory, this book analysed how different interests regarding fuel economy regulations and state-automobile industry

M. Iguchi, *Divergence and Convergence of Automobile Fuel Economy Regulations*,
DOI 10.1007/978-3-319-17500-3_7

relationships have developed, how they have interacted, and how these interactions have resulted in converging fuel economy standards.

To do so, I set following three hypothesis that explain the logic of the converged fuel economy regulations: the first concerns an assumption that governments' effort to harmonize national policies through negotiations caused the regulatory convergence; the second hypothesis draws on what Vogel argued the 'California effect', namely global competitive pressures that drive regulatory competition to adjust their national policies; the third hypothesis sketches voluntary adjustments by political actors through policy diffusion and learning.

According to interviews conducted to various stakeholders in Europe, Japan, and the US, it was clear that there were no obvious inter-governmental coordination to regulate CO_2 emissions from road transport sector between Europe, Japan and the US. Rather, case studies tell us that intergovernmental coordination to harmonize fuel economy regulation has been difficult to achieve due to following two reasons. First, climate policies in the road transport sector take varieties of national contexts into account. This includes: the geographic features (structures of cities and distances between rural areas, including availability of public transport, and regional differences in travelling coefficients); consumer preferences and differences between industries' traditions/cultures or identities with more than 100-year histories (e.g. consumer preferences for domestic brands—bigger vehicles are more likely to be sold in the US, whereas Japanese and European manufactures produce relatively small vehicles); and the availability of oil resources and a price of fuel (the US is one of the major oil producing countries but Japan and Europe depend on foreign imported oils). All of these factors make a coordinating a simple global unified policy for the road transport sector difficult to achieve. Therefore, the first hypothesis was not confirmed.

Regarding the third hypothesis, case studies show that international policy networks of environmental NGOs and/or business did not emerge to encourage policy diffusions and learning among political actors. Interviews from key stakeholders proved that there were no obvious coordination among NGOs nor businesses in three countries/region. European environmental NGOs have been engaging specifically with the process of European fuel economy regulations. Likewise, American environmental NGOs have only been active within the US. Furthermore, Japanese environmental NGOs were not influential in the process of Japanese fuel economy regulations. According to interviews conducted to various environmental NGOs in Europe and the US, the fundamental reason of such absence is rooted in regional varieties of car climate policies. For instance, while revealing low achievements of the industry to fulfil the voluntary agreements standards were effective in Europe, this strategy would not bring any effects to the US auto industry since the CAFE regulation itself has already been stagnant. As for the industry coordination, there have been no incentives for them to harmonize regulatory standards. According to interviews conducted to various companies and business networks in the three studied case studies, industry harmonization towards a single fuel economy standard may cause clash of interests since they are operating in different test cycles (e.g. 'CO_2/km' in Europe, 'L/km' in Japan and 'mpg' in the

US). To put simply, such harmonization would impose significant costs to other manufactures to re-introduce the new test cycle. Even though the third hypothesis was not confirmed, it should also be noted that the role of non-state actors have been increasingly important. For example, environmental NGOs played a critical role in the EU fuel economy regulations and California was instrumental to push for stringent federal level regulations in the US. Although networks among these actors have not emerged yet, the regulatory convergence may encourage future developments of such networks by eliminating regional varieties of car climate policies and provide incentives for business actors to coordinate their strategies.

Hence, this book confirmed what really triggered the convergence of fuel economy regulations were endeavours of governments to enhance its competitiveness of the automobile industry by raising fuel economy regulations, in reactions to global competitive pressures.

Automobile industry transformed from the position of 'dragger' to 'pusher' towards solution of climate change issue through business conflicts over stringent fuel economy regulations. Business conflicts are motivated to create a global or regional level playing field in order to have competitive advantage over competitor firms. This logic shares a lot of characteristics with what Vogel (1997) called as the 'California effect'. The argument here is that the California effect takes place when a country impose stricter standards upon one or more if its trading partners through the use of market access. Foreign companies have to comply with the stricter standards if they wish to continue to export their products; otherwise, they may face competitive disadvantages over rivals based on stricter standard because they may have to maintain separate production lines for export and domestic market. It may be therefore an advantage for foreign companies to support stricter standards in their home countries.

I conclude that this logic of competition was the central driving force for automobile industry to support for stringent fuel economy regulations. The result was the regulatory convergence of fuel economy regulations that is born out from regulatory competition among the major automobile manufacturing nations with the rationale to enhance its competitiveness of the automobile industry. By revealing this mechanism, this book confirmed the validity of 'business conflict approach' (Falkner 2001, 2003, 2005, 2008) to explain the recent trend of rising fuel economy regulations among major automobile manufacturing countries and the region. In other words, business competitions were key determining factor that led to the regulatory convergence between Europe, Japan and the US (see Chap. 6 for details).

7.2 Theoretical Contributions

The above section answered the research questions set in Introduction. Based on these findings, this section emphasises the academic contributions that this book made within the discipline of the IR. It firstly points out how this study advanced the existing studies mainly done by the Variety of Capitalism approach. Then, it extracts generalisation drawn from studied case studies and explains how this study contributed to the constructivist theory of IR.

Beyond the 'Variety of Capitalism'

Varieties of Capitalism (VOC) approach assumes different institutional structures according to different countries, especially the relationships between government and industry and how they influence different patterns of innovation in each country into following two types: the Coordinated Market Economy (CME) such as Europe and Japan; and the US as the Liberal Market Economy (CME) (Hall and Soskice 2001). Mikler (2006, 2007, 2009, 2010) argues that the CME countries are likely to have higher standards of fuel economy regulation, while, he explained, US fuel economy regulation has been stagnant because of its LME tradition.

While VOC approach has greatly advanced our understandings on the relationships between different types of capitalist system and environmental standards, however, since its ultimate focus is placed on the *institutional arrangements* of each country, it fails to account *dynamics of actor relationships* that actually triggered the convergence of fuel economy regulations.

With these findings, this book enhanced explanatory power of the VOC approach in following regard: it not only enabled to analyse fuel economy regulations in advanced industrial states, but also enabled to observe the patterns of fuel economy regulations in emerging economies. Namely, the VOC approach that combines insights of Constructivist theory could enrich political development theory that attempts to observe future directions of environmental policies in developing countries by looking at the development patterns of environmental policy in advanced industrial states.

Drawing upon the findings that are shown in the Table 6.1, this book makes following three classifications of development patterns of fuel economy regulations that emerging economies could follow. Firstly, in the EU type of supranational state that involves various actors with divergent interests, the climate driven norms could stir clashing interests and promote stringent regulations. This norm could be produced by a committed supranational political body such as the European Commission as well as environmental NGOs. The norm generated by these political actors would be embedded in the decision-making process through co-decision procedures between related supranational political bodies. Once the norm is embedded in the decision-making process, the numerical targets are 'locked-in' and would be hard to remove. The industry would then compete to reflect their preferences on how to achieve the target, and this business conflict could push for further regulatory standards, as Germany proposed the 120 g/km target and France proposed the 95 g/km target. Hence in this model, the role of norm entrepreneurs and the 'sticky' decision-making process, along with business conflicts between the industries within the supranational market could foster the stringent fuel economy regulations.

Secondly, a country that has a tradition of coordinated decision-making process between the government and the industry, with high dependence on foreign oil and a relatively small size of its domestic market could follow Japanese pattern of fuel economy regulations. In this type, enhancement of fuel efficiency of cars are said to be critical, since it co-benefits to energy-savings and to enhance its industry competitiveness in the international market. Such regulations could be enabled by decision-making processes based on the co-regulation that involves rather limited actors whose interests and norms are coordinated by their own networks. Hence in this model, the factors including coordinated norms and interests between the government and the industry, along with enhancement of domestic industry competitiveness in overseas markets would be the key to promote the stringent fuel economy regulations.

Thirdly, a country that has a tradition of liberal market economy based on free competitions among industries with relatively large domestic market could follow the US model. Critical changes can be brought by a powerful norm entrepreneur such as the state of California. California played a critical role in the US fuel economy regulations with its large market power and proactive state-level environmental regulations that are strengthened by the EPA under the Obama administration. Its own regulatory standards changed the industry behaviour to support improvements of the CAFE regulations. Hence, in this model, a proactive local government under the rule of pro-environment central government could serve as the norm entrepreneur that could change whole political landscapes in liberal market economy. In turn, domestic car industry and related stakeholders are likely to accept stringent fuel economy regulations in order to enhance its competitiveness over foreign car industry that operates in the domestic market.

Constructing Constructivist Theory: The Regulatory Convergence Is What Competition Make of It

In the Chap. 2, 'Analytical Framework', this book showed validity and prospects of constructivist theory of IR to analyse the regulatory convergence of fuel economy regulations between Europe, Japan, and the US. In particular, this book takes the 'Holistic constructivism' that emphasises mutual constitutive nature between international and domestic politics, and seeks to bring them together into a unified analytical perspective. Figure 7.1 shows the image of holistic constructivist model in the discipline of IR.

This model emphasises that interest of agent (i.e. state) is not given, but socially constructed through interactions with other agents. These interactions in turn, are also reproduced and consolidated into norms and institutions and therefore become structure. In turn, the structure further re-constructs the interests of states, and hence

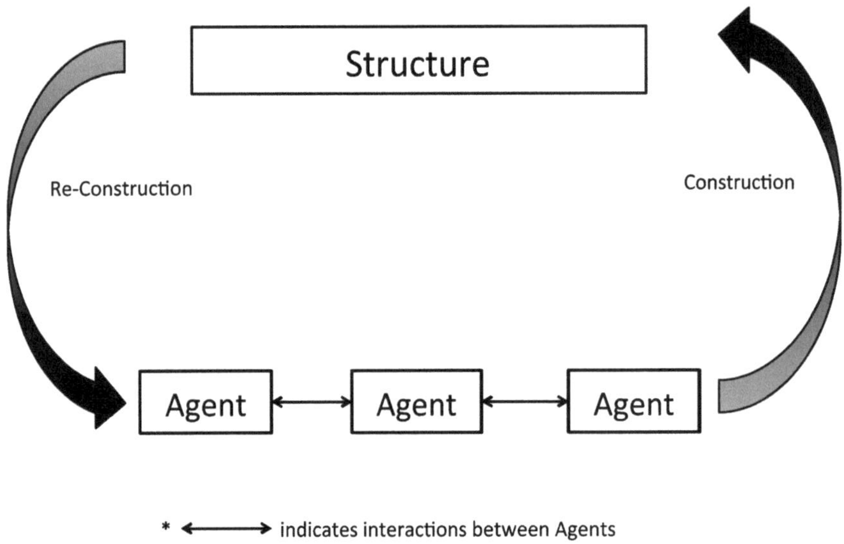

Fig. 7.1 Holistic constructivist model. *Source* created by author

cultivates the behaviour of states. In other words, the interest of state does not solely exist out there in international politics; nor state interest is only be defined by any sorts of international structure; but it is the result of interactions between various political actors, including government, business and environmental NGOs.

As applied to observe the regulatory convergence, I have mentioned that the theory helps us to reveal the patterns of state-automobile industry relationships that develop differently according to the context of each state and the region, as well as to explain why the international competition on fuel economy regulations were emerged and how it re-constructed fuel economy regulations in other countries that directed towards the regulatory convergence.

Figure 7.2 shows the application of Holistic constructivist model to observe the regulatory convergence of car fuel economy between Europe, Japan and the US. In this model, the interest of each agent is defined through interactions through competition. Such competitions among states are backed up by business competitions among auto industry in each country/region. Business competition arise when auto industry operate its economic activities in other country face stricter environmental standards imposed by one or more export countries. In order to stay competitive in both foreign and domestic markets, auto industry is likely to support stringent standards in their home countries.

Japanese case fits well in the description above. Japanese fuel economy regulations were largely cultivated by European and to extent California's fuel economy regulations, since both markets were crucial for Japanese automobile industry. In particular, Japanese fuel economy regulations for 2015 (125 g/km) and 2020 (105 g/km) are set in line with the European level (130 g/km by 2015 and 95 g/km

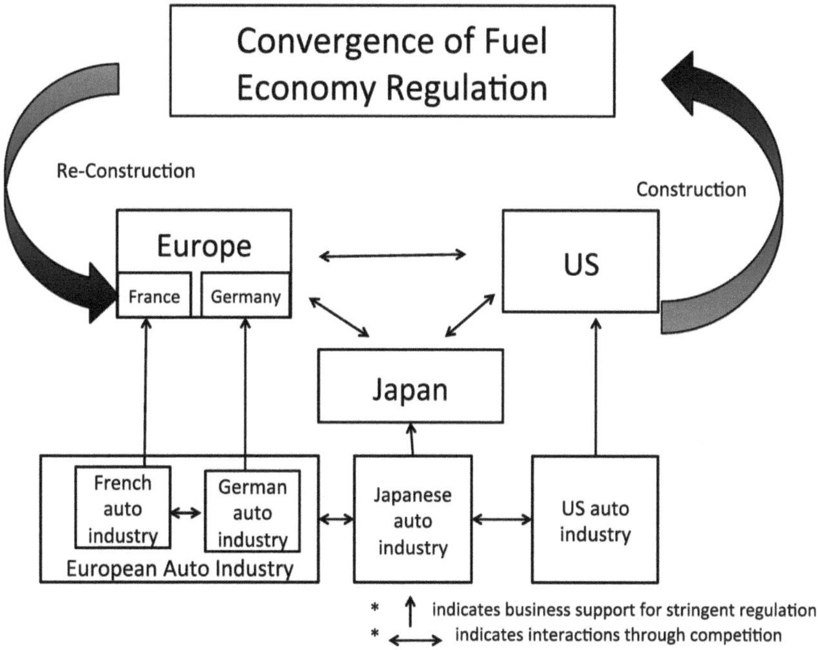

Fig. 7.2 Business actor driven model of holistic constructivist model. *Source* created by author

by 2020). Furthermore, Japanese automobile industry showed its business support for stringent fuel economy regulations in domestic market rather cooperatively through co-regulations between its governments.

In the case of the US, automobile industry supported stringent fuel economy regulations up to the level of European and Japanese standards, due to the loss of its competitiveness within the domestic market where they gained most of profits. However, such support was not voluntary in nature; rather, it was triggered by the state of California's own emissions standard introduced in response to the Supreme Court decision in 2007, because the US automobile industry faced competitive disadvantages over rivals based on stricter standard because they may had to maintain separate production lines for California and other states.

The European fuel economy regulation was driven by business competition between French and German automobile industries over supranational standard settings. The stringent target was proposed by Germany with the aim to introduce weight-based regulation that would benefit the German industry; on the contrary, France claimed for stringent but a regulation based on an absolute numbers (i.e. the target that applies to all companies, no matter heavy the cars they produce). The proposed targets born out from such competitions are then legalized by the European supranational decision-making process based on the codecision procedure.

These findings and the application of holistic constructivist model to observe the regulatory convergence of car fuel economy between Europe, Japan, and the US advanced the applicability and our understanding of constructivist theory in following two fronts.

First contribution is that it enriched the explanatory power of constructivist theory. Indeed, the birth of constructivist theory in 1980s broadened the new research area of international politics—that is, the importance of ideational forces (ideas, cultures, norms and socialization) in contrast to material forces (for example, a numbers of nuclear weapons) in defying the behaviour of states. The good example would be the statement made by Alexander Wendt in 1995. He said: "500 British nuclear weapons are less threatening to the United States than 5 North Korean nuclear weapons" (Wendt 1995: p. 73). However, less well known are about how such ideational forces emerge, by whom, when and how it forms interests of actors in any given studied area, due to little evidences to support its claims (Mearsheimer 1994, pp. 44–47). By revealing the mechanism of the regulatory convergence of car fuel economy by using the holistic constructivist theory with concrete evidences collected through semi-formal interviews to key stakeholders, this book added one but important case study that proved the explanatory power of the constructivist theory.

The second contribution is, this book enabled to explain the mechanism of regulatory convergence of car fuel economy by combining constructivist theory and business actor approach. While systematic constructivists such as Wendt adopts 'unitary state actor' assumptions and ignores the importance of domestic political actors, this book demonstrated business competitions are the key determinants in defying state behaviours over stringent car fuel economy standards. It has also pointed out that various political actors including environmental NGOs and local state actors were instrumental to change the behaviour of businesses. Hence, this book displayed that not only interactions between states, but also interactions between non-state actors in constructing the state interest are critical factors that we cannot simply ignore for the future development of constructivist theory.

7.3 Applicability of 'Agency Within and Beyond the State' Model of Regulatory Convergence to Environmental Policies

The constructivist model of regulatory convergence of environmental policies that developed based on the case studied in this book could apply to business actors that have following characteristics:

Multinational corporations that are sensitive to environmental standards of export markets;

Second, improvements of environmental quality of their products (such as higher energy efficiency or reducing hazardous substances) co-benefits to an enhancement of its competitiveness against their rivals. Significant improvements of environmental policies in the large market would in turn encourage international convergence of the stringent regulatory standards.

Electronics industry would fall into such category. In fact, environmental standards to regulate hazardous substances in electrical and electronic products (such as Personal Computers, cell phones, and televisions) are converging between major manufacturing countries with the introduction of EU's Directive on the restriction of the use of certain Hazardous Substances in electrical equipment (RoHS) adopted in 2003.[1]

The regulatory standards set by the RoHS were then adopted by other countries, such as in California (Electronic Waste Recycling Act of 2003),[2] in Korea (the Act for Resource Recycling of Electrical and Electronic Equipment and Vehicles in 2008) and in China (Administrative Measure on the Control of Pollution Caused by Electronic Information Products in 2009) (METI n.d.). Japan does not have regulation to restrict hazardous substances from electronic products; however, the amendment of 'Law for the Promotion of Effective Utilization of Resources' in 2006 and a ministerial ordinance Japanese industrial standard for Making of Specific Chemical Substances (J-MOSS) in 2006 requires any electrical and electronic products containing hazardous substances exceeds the standard value equal to the RoHS to indicate on the product itself, the packing, and on catalogues and other documentation (JEITA 2008).

It follows, environmental regulations on electronics industry are converging across major markets. Although an applicability of the constructivist model on this issue are subject of further investigation, Bidenkopf (2012) points out that Chinese adoption of the RoHS standards was triggered by industry competition, in combination with domestic environmental problems. This finding implies that the constructivist model developed in this book that emphasize enhancement of industry competitiveness as the key driving forces to encourage higher environmental standards, could apply to the regulatory convergence of hazardous substances on electrical and electronic products.

Having said the theoretical contributions of the book, next section seeks empirical implications that the regulatory convergence would place in considering the future global climate governance.

---

[1]EU's RoHS took effect in 2006 and restricts the use of following six substances: Lead, Mercury, Cadmium, Hexavalent chromium, Polybrominated biphenyls, and Polybrominated diphenyl ether. The maximum permitted concentrations in each homogenous material in products are 0.1 % (except 0.01 for Cadmium) by weight. EU had also adopted Waste and Electronic Equipment Directive (WEEE) in 2003, which set collection, recycling and recovery targets for all types of electrical goods, as well as Registration, Evaluation, Authorisation and Restriction of Chemical (REACH) in 2006, which addresses assessment and management of the risks posed by chemicals and provide appropriate safety information to consumers.

[2]Applies to following four substances: Lead, Mercury, Cadmium, and Hexavalent chromium. The maximum permitted concentrations in each products are 0.1 and 0.01 % for Cadmium (California State Board of Equalization 2007).

7.4 Future Tasks

On the final note, the next step of the research is threefold. First, the effectiveness of the de facto standard in influencing the behaviour of the emerging economies must be measured. In other words, while this book focused on the case studies from Europe, Japan and the US, more case studies, particularly from China and India, are needed to explore on how the regulatory convergence of fuel economy in the world's biggest car markets would place impacts on fuel economy regulations of other countries.

Second, more micro level analysis on strategies of automobile industry is needed to advance this study. The primal focus of this book is placed on the convergence of fuel economy regulations, by looking at political dynamics of each studied case studies that pushed automobile industry from the position of the dragger to the pusher. Therefore, this is the study about institutions and political actors in the context of climate change issue. However, if we are to investigate more on the political dynamics, an interdisciplinary approach to merge more micro level analysis on the strategies of automobile industry in each country in relation to technology structure and their innovation management is needed for further analysis (Fujimura et al. 2012).

Lastly, the findings of this book could potentially be related to the future architecture of the low-carbon technology governance (Kanie et al. 2013). If the 'race to the top' of car fuel economy regulations is happening, this means that the industry competitions over sustainable technological innovations and its resulted 'de-facto' standard could be one of the effective mechanisms within the 'fragmented' climate change governance (Biermann et al. 2009). Thus, the next task is to further explore how business conflicts over sustainable technology innovation could foster the convergence of environmental regulations in countries at stake, as well as how the converged regulatory standards can effectively work as one of the mechanisms in global environmental governance.

References

Biermann F et al (2009) The fragmentation of global governance architectures: a framework for analysis. Global Environ Polit 9(4):14–40

Bidenkopf K (2012) Hazardous substances in electronics: the effects of European Union risk regulation on China. Eur J Risk Regul 3(4):477–488

California State Board of Equalization (2007) California electronic waste recycling Act of 2003, California

Environmental Protection Agency (2011) Federal register, vol 67, No 179. Washington DC

Falkner R (2001) Business conflict and US policy. In: Harris PG (ed) The environment, international relations, and U.S. foreign policy. Georgetown University Press, Washington, DC, pp 157–177

Falkner R (2003) Private environmental governance and international relations: exploring the links. Global Environ Polit 3(2):72–87

Falkner R (2005) The business of ozone layer protection: corporate power in regime evolution. In Levy DL, Newell PJ (eds) The business of global environmental governance. MIT Press, Cambridge, MA, pp 105–134

Falkner R (2008) Business power and conflict in international Environmental politics. Palgrave McMillan, Basingstoke

Fujimura et al (2012) The necessity of inter-industry mediation for harmonized industrial progress. In: Nilsson M, Hillman K, Rickne A, Magnusson T (eds) Paving the road to sustainable transport governance and innovation in low-carbon vehicles. Routledge, London, pp 117–135

Hall PA, Soskice DW (2001) Varieties of capitalism: the institutional foundations of comparative advantage. Oxford University Press, Oxford

JEITA (Japan Electronics and Information Technology Industries Association) (2008) About J-MOSS. http://home.jeita.or.jp/eps/epsJmoss.html. Accessed 17 Dec 2013

Kanie N, Suzuki M, Iguchi M (2013) Fragmentation of international low-carbon technology governance: an assessment in terms of barriers to technology development. Global Environ Res 17(1):61–70

Mearsheimer JJ (1994) The false promise of international institutions. Int Secur 19(3):5–49

Mikler J (2006) Varieties of capitalism: national institutional explanations of environmental product developments in the car industry. The University of Sydney

Mikler J (2007) Varieties of capitalism and the auto industry's environmental initiatives: institutional explanations for firms' motivations. Bus Polit 9(1):1–38

Mikler J (2009) Greening the car industry: varieties of capitalism and climate change. Edward Elgar, Cheltenham

Mikler J (2010) Business as usual or apocalypse now? Regulating the car industry's carbon emissions. In: Third biennial conference of the European consortium on political research standing group on regulatory governance. Dublin

MLIT (n.d.) CO_2 emissions from transport sector. http://www.mlit.go.jp/sogoseisaku/environment/sosei_environment_tk_000007.html. Accessed 4 Sep 2013

Transportpolicy.net (n.d.) Japan: heavy-duty: fuel economy. http://transportpolicy.net/index.php?title=Japan:_Heavy-duty:_Fuel_Economy. Accessed 4 Sept 2013

Vogel D (1997) Trading up and governing across: transnational governance and environmental protection. J Eur Publ Pol 4(4):556–571

Wendt A (1995) Constructing international politics. International Security 20:71–81

Index

M. Iguchi, *Divergence and Convergence of Automobile Fuel Economy Regulations*,
DOI 10.1007/978-3-319-17500-3

MIX
Papier aus verantwortungsvollen Quellen
Paper from responsible sources
FSC® C105338

If you have any concerns about our products,
you can contact us on
ProductSafety@springernature.com

In case Publisher is established outside the EU,
the EU authorized representative is:
Springer Nature Customer Service Center GmbH
Europaplatz 3, 69115 Heidelberg, Germany

Printed by Libri Plureos GmbH
in Hamburg, Germany